HYDRAULIC POWER
SYSTEM ANALYSIS

HYDRAULIC POWER SYSTEM ANALYSIS

Arthur Akers
Iowa State University
Ames, Iowa, U.S.A.

Max Gassman
Iowa State University
Ames, Iowa, U.S.A.

Richard Smith
Iowa State University
Ames, Iowa, U.S.A.

Taylor & Francis
Taylor & Francis Group
Boca Raton   London   New York

A CRC title, part of the Taylor & Francis imprint, a member of the
Taylor & Francis Group, the academic division of T&F Informa plc.

Published in 2006 by
CRC Press
Taylor & Francis Group
6000 Broken Sound Parkway NW, Suite 300
Boca Raton, FL 33487-2742

International Standard Book Number-10: 0-8247-9956-9 (Hardcover)
International Standard Book Number-13: 978-0-8247-9956-4 (Hardcover)

Library of Congress Cataloging-in-Publication Data

Catalog record is available from the Library of Congress

Visit the Taylor & Francis Web site at
http://www.taylorandfrancis.com

and the CRC Press Web site at
http://www.crcpress.com

Preface

The text introduces and reinforces key principles, concepts, and methods of analysis of the performance of fluid power components and systems. The physical configuration of individual components is presented and this information is supplemented with material relating to dynamic analysis. The principles of analysis have been demonstrated with a comprehensive range of worked examples and with suitable exercises for the student to follow in order to acquire considerable command of fluid power system design details.

Some of the ways are shown where fluid power can be used to advantage in engineering systems. Fluid power may often provide a way to transmit power that, for certain levels of power and other circumstances, may be superior to other mechanical or electrical techniques.

The text has been written primarily for mechanical, aerospace or agricultural engineering seniors or for graduate students undertaking research to extend the limits of fluid power technology. The text should also be helpful for engineers working on design or research projects in manufacturing facilities who require knowledge of computer simulation of the dynamic performance of fluid power components and systems.

We now take the opportunity to express our appreciation for the efforts of all who brought this text to fruition. We acknowledge gratefully the advice provided by the prepublication reviewers: Richard Burton, University of Saskatchewan, and several others. Our consultations with Brian Steward of the Agricultural and Biosystems Engineering Department of ISU have provided us with excellent and original ideas to help us with our text.

We wish to thank the publisher for their trust in us. This text was started under the guidance of Steve Sidore and John Corrigan of Marcel Dekker. The content of the text owes much to these two individuals whose knowledge of the fluid power field was extensive. After Marcel Dekker was assimilated by Taylor & Francis, our editors there were Jessica Vakili and Jay Margolis. We would like to thank Jessica for her help and patience as deadlines were agreed upon and missed. We should also like to thank Jay for his painstaking attention to detail and his ability to find errors that we had overlooked even after countless readings of the text.

Producing a book is not a trivial task, but is somewhat simplified by the availability of computers and the programs that run on them. This text was produced using LaTeX. Any author using LaTeX owes a huge debt of gratitude to the many members of the TeX Users Group who maintain and enhance LaTeX. Production of this text was very much facilitated by several packages developed by the many people associated with TUG.

Lastly we owe a debt of gratitude to our wives, namely Marcia L. Akers, and Gail E. Gassman. Without their care, their help, encouragement and patience, this work could never have been completed.

A. Akers
M. P. Gassman
R. J. Smith
Ames, Iowa

Chapter Synopsis

Chapter 1 gives a description of how fluid power is used, a brief history of fluid power activities to the present, some projections for future applications, and some advantages of using fluid power for mechanical power transmission.

Chapter 2 outlines details of the U.S. Customary and (the more modern) S.I. Systems of units. Conversion between these two systems is discussed, and the worked examples switch randomly between the two systems throughout the text. The physical properties of fluids described are oil density, viscosity, bulk modulus, specific heat and thermal conductivity. Values of these properties are usually required to design the components of fluid power systems.

Chapter 3 outlines procedures for steady state modeling. In this text, steady state is defined somewhat freely as a condition where a system may change state with time. The changes with respect to time, however, are *sufficiently slow* that algebraic equations rather than time dependent differential equations may be used for the problem solution.

Sources of the mathematical equations used are given where the principles described are conservation of flow and of energy. The main part of the chapter deals with conversion of pressure energy into heat energy in various forms of fluid flow in pump and motor systems. A model of a flow regulator valve is presented showing how preliminary estimates of valve opening and flow can be made using steady state analysis. The chapter also includes an example of using an accumulator in a system to reduce pump size and energy use.

Chapter 4 gives the development of analytical methods for determining the dynamic behavior of fluid power systems. In order to do this, Newton's Second Law of motion is invoked together with the phenomenon of pressure change as a function of volume change affected by fluid bulk modulus.

Thus it is shown that the dynamic performance of fluid power components and systems can be described by sets of ordinary differential equations with displacements, velocities, and pressures as the state variables. A servovalve controlled actuator and the same system with positional feedback are presented as worked examples. As noted in the text, there are numerous pieces of software and programs in the technical literature that may be used to solve these sets of equations numerically.

Chapter 5 gives a brief review of the Laplace transform method of solving linear ordinary differential equations. The material is presented as a precursor to discussion of stability, the spring-mass-damper, and the concept of a time constant. The Laplace transform approach also gives an opportunity to introduce the block diagram, which is often a very helpful intermediate step between a physical system and its representation as a set of differential equations. The chapter also discusses the consolidation of block diagrams and the concept of a transfer function.

Chapter 6 shows how the consolidated transfer function developed in Chapter 5 may be used to predict the steady state response of a linear system to a constant sinusoidal input excitation. The chapter then discusses ways of using frequency response for establishing an appropriate level of controller gain in a feedback system. Although there are a number of procedures to select feedback gain to produce a stable system, the procedure using frequency response is felt to be particularly applicable to fluid power systems.

The hydromechanical servo examined in Chapter 4 is revisited and controller gain is determined using the frequency response method. The text also indicates how frequency response diagrams can be generated from a mathematical model of the system and indicates that the diagrams may also be generated experimentally.

Chapter 7 deals with valves, the purpose of which is to modulate flow rate, direction of flow, flow sequence, and to control pressure. The text is provided with many figures showing how these tasks are accomplished.

The spool valve is commonly used for proportional control. Modulation of large spool valves by solenoids and torque motors may be difficult because of the flow forces developed by fluid momentum changes in the valve. Expressions for the magnitude of these flow forces are developed. Linearization of the characteristics of spool valves is developed because these expressions are required when mathematical models of valves are used in the analysis of automatic control systems.

Chapter 8 outlines the design details of pumps and motors. There are hydrodynamic pumps which generate rotation of the fluid. This rotational kinetic energy is transformed into pressure. Such pumps are seldom used in fluid power applications for which constant outlet pressure is required. The pumps usually employed are known as positive displacement or hydrostatic pumps.

There are seven types described in the text and it is demonstrated which type is optimum for certain applications. Motoring action may be achieved by pumping oil through a hydrostatic unit. In some instances the geometry of a pumps differs little from that of a motor. Such a situation may not always be valid.

It is possible to obtain values of efficiencies from manufacturers' catalogs for different speeds and pressures so that a correct choice can be made for each application.

Chapter 9 deals with axial piston pumps, which are the type most frequently used where precision and high volumetric efficiency are required. The type of axial pump described is the swash plate design and pressure transition during pump rotation is analyzed. It is demonstrated how the pressure variation is affected by the geometry of the fluid outlet and it is shown how a suitable compromise between controlling pressure rise and maintaining volumetric efficiency may be achieved. The chapter ends with a discussion of the variation of torque and flow rate associated with multiple piston pumps.

Chapter 10 discusses the operation of hydrostatic transmissions. A conceptual design for a mechanical hoist is presented that has a gearbox and a progressive action clutch that enables the speed to be varied economically and smoothly. It is suggested that such design would not be suitable for a hoist because of the intermittent nature of the power transmission. The example was used to show the need for a stepless transmission. Other stepless transmissions using sheaves or electric motors are briefly described in the text. The usual components of a hydrostatic transmission are a variable displacement pump and a fixed displacement motor.

A typical plot of motor torque and motor speed shows that the transmission can be limited for torque at low output speeds and by power at high output speeds. A typical performance envelope is presented. The design of a hydrostatic transmission to drive a research soil bin is described. The analysis is presented and a block diagram of the governing equations is given. The pressure and velocity of the bin have been presented. The velocity profile was shown to be greatly affected by the value of bulk modulus chosen.

Chapter 11 describes a pressure regulating valve, an essential device in a hydraulic system because the supply pump is a positive displacement device (see Chapter 8). Without such a valve, the pressure would increase until damage and failure of machine parts would occur. Dynamic operation of the valve is described and the governing equations of motion are developed. Providing a solution to the challenging problem of valve spool damping is also discussed.

Chapter 12 extends the model given in the previous chapter. An actuator and load are added and the equations of motion of the dynamic behavior of the complete system are developed. Computer simulation is a tool that allows the geometry of the various parts to be modified to satisfy the particular needs of the system.

Chapter 13 presents flow division using a modification of Ohm's law for electrical resistance, $\sqrt{\Delta p} = RQ$. Expressions for consolidating series and parallel sets of resistances are presented for this law. Worked examples are presented of increasing complexity, culminating in the dynamic analysis of a flow regulator valve.

The limiting case of steady flow was presented by Esposito [1][1]. The $\sqrt{\Delta p} = RQ$ law approach was considerably expanded by Gassman [4-6] to solve problems with time variable flow conditions and orifice areas. The general problem of the dynamic analysis of a flow regulator valve entails bifurcated flow with multiple varying resistances. The solution is achieved by an iterative procedure that automatically updates the varying resistances, while concurrently solving the dynamic equations in displacement, velocity, and pressure using a conventional differential equation solver. The authors believe that this is the first appearance of the solution in the technical literature.

Chapter 14 shows that positive displacement pump design employs pistons, vanes or gear teeth that cause the oil flow to pulsate. The resulting pressure waves in the fluid stream cause vibration noise at frequencies related to the rate of pumping or pump rotation. It is shown that the analogy between electrical and fluid circuit resistance introduced in Chapter 13 may be extended for the properties of capacitance and inductance. Fluid circuit expressions for inductance are developed for plug flow and laminar flow in a circular pipe. An example is given of applying the electrical/fluid analogy to noise reduction for a tractor hydraulic pump.

---

[1]Reference numbers in this discussion of Chapter 13 refer to the references at the end of that chapter.

Programs

INTRODUCTION

A CD of programs accompanies this text. It is the authors' belief that a new comer to fluid power analysis should start by writing equations, solve those equations with a general purpose programming language, and display the results in the form of tables and graphs. Only when the reader has become experienced with the formulation of equations and discovered that the analysis of systems requires assumptions and simplifications, should he/she graduate to more complex application packages that often conceal these factors.

The spreadsheet program Excel® is a very good general purpose calculating program that allows easy plotting of results. Many users are familiar with the cability of placing formulae in cells. Although this is useful capability, its utility becomes limited as the complexity of the task increases.

Many users know that Excel® can record macros from a series of key strokes or mouse clicks. These actions are translated by Excel® into a Visual Basic for Applications® (hereafter VB) subroutine that can be called by the user. It will be seen that VB may be used to write quite complex programs.

The major reason for the Excel® programs that are being provided here is that many people who have a computer also have this program installed. Initially, the reader can try out analysis of fluid power systems without needing to invest in other mathematical programs such as Mathcad® or MATLAB®. It is not suggested that Excel® replaces such mathematical programs for serious design work, but a student can use the program to gain familiarity with analytical concepts covered in the text before moving on to the more powerful tools.

This text cannot pretend to be a manual exploring all the capabilities of VB. It is hoped, however, that the material presented in the readme.pdf file in the root directory of the disk will allow the reader to start writing VB programs.

EXCEL PROGRAMS

Chap4\ch4_ex2.xls

The program simulates the operation of a walking beam feedback unit. It turns out that this is a stiff system and cannot be integrated with a Runge Kutta 4th order adaptive stepsize routine. Two backward Euler routines, Bulirsch Stoer and Kaps Rentrop, translated from C++ routines given in Numerical Recipes 2nd ed. [1] are also provided and the user can choose any one of the three routines. The RK4 routine fails because the stepsize becomes vanishingly small. The BS and KR routines are both capable of integrating the differential equation set.

chap6\frq_rsp.xls

The program generates amplitude ratio vs. frequency and phase angle vs. frequency (Bode plots) for the ratio of two polynomials. Automatic chart generation is provide. The number of polynomial coefficients and the values of the coefficients may be changed as desired.

The program introduces a new VB feature. A programmer can define his/her own VB functions. This capability has been used to define a complex variable, which is missing in VB. Other subroutines are provided that perform complex arithmetic.

chap9\ptrans_smpl.xls

This program performs a very simplified analysis of the pressure experienced in the cylinder of an axial piston pump as it passes across top dead center. There is only one equation, that in pressure, so the problem can be solved using a non stiff Runge Kutta method.

The reader should examine the VB code in conjunction with Figure 9.2 to see the meanings of the variables. Several figures in Chapter 9 were produced using this program.

chap10\hydrsttrans.xls

This program simulates a soil bin being accelerated up to a constant speed, being held at constant speed during tool engagement, and being decelerated. The program is fairly user friendly and its structure resembles chap4\ch4_ex2.xls. There are routines for writing a sample data set and plotting is performed automatically. Although there are 3 state variables, the system was not stiff enough to cause trouble integrating with a fixed step size Runge Kutta.

The data provided are for the initial design using flexible hoses that had a large contained oil volume and a low effective bulk modulus. The reader may wish to try altering the oil volume and bulk modulus to see what improvement can be made in tracking the desired velocity conditions.

MATHCAD AND MATLAB PROGRAMS

chap11\section11_2.mcd

This file solves the pressure relief valve problem in Chapter 11 using Mathcad®.

chap11\section11_2.mdl

This file is an alternative solution for the relief valve using MATLAB®.

chap13\chap13.mcd

This file solves the flow divider valve problem in Chapter 13 using Mathcad®.

chap14\chap14.mcd

This file solves the tractor pump attenuator problem in Chapter 14 using Mathcad®.

chap14\chap14.mdl

This file is an alternative solution for the tractor pump attenuator problem using MATLAB®.

COMMENTS

None of the examples are for very complicated systems. On the other hand, the reader should be able to extend the examples to more state variables as the system being analyzed becomes more complex.

1. Press, W. H., Flannery, B. P., Teukolsky, S. A., Vetterling, W. T., 1992, *Numerical Recipes The Art of Scientific Computing*, 2nd ed., Cambridge University Press, Cambridge, U.K.

Contents

CONTENTS

Figures

Tables

Glossary of Variables

Comments

- Some quantities, such as diameter, d, radius, r, length, ℓ, force, F, etc., are used in the many different expositions throughout the text. It was felt that attempting to include the combination of every variable and its identifying subscript would not be helpful. On the other hand, most well established combinations, such as C_d for the coefficient of discharge for an orifice or K_c for the flow gain coefficient for a spool valve, have been included as their explicit combinations.

- To conform to the above statement, where the same symbol occurs with several subscripts, e.g., d with $_a$ for actuator or $_v$ for valve, only the unsubscripted symbol is listed in this glossary. The subscripts should be obvious in context and, in most instances, subscripts are defined where they are used.

- A text relating to fluid power draws from several disciplines and each discipline has traditionally used certain symbols. Rather than introduce alien symbols, some chapters may use the same symbol for more than one characteristic. In this glossary, such symbols have been annotated with their chapter number.

- Dot notation for differentiation with respect to time is used in the text, but not recorded for each time varying symbol in the glossary.

- The symbols F, M, L, T, and Θ are used for the primary dimensions of Force, Mass, Length, Time, and Temperature.

- The lower case symbols i, m, n, and r are used in several locations as counting variables in sequences. These lower case symbols are used with other meanings, e.g., i for electrical current. The different meanings are believed to be obvious in context even when the symbols appear in the same block of text.

LOWER CASE LATIN

a	Acceleration
a	Reciprocal of time constant for 1st order system, Chapter 5
a_j	Denominator factor in partial fraction expansion of $N(s)/D(s)$
c	Coefficient of viscous drag
d	Diameter
e	Base of natural logarithms (2.7183)
f	Fraction, Chapter 2
f	Frequency (Hertz), Chapter 4
f	Head loss friction factor, Chapter 3, Chapter 13
g	Acceleration due to gravity
g_c	Constant relating mass units, Chapter 3
h	Head loss as height of fluid (pipe flow)
i	Electrical current
j	$\sqrt{-1}$
k	Spring rate
ℓ	Physical linear dimension
m	Mass
n	Index of gas polytropic expansion or compression, Chapter 3
n	Number of teeth, Chapter 8
n	Root multiplicity in partial fraction expansion of $N(s)/D(s)$, Chapter 5
n	Rotational speed (revolutions per minute), Chapter 3
p	Pressure
r	Amplitude of a complex number expressed in polar form (a real quantity), Chapter 6

Continued on next page

Continued from previous page

r	Radius
s	Laplace domain independent variable
t	Time
$u(t)$	Unit step input
$u(0)$	Unit impulse input
v	Velocity
w	Flow area gradient
x	Linear displacement
x	Abscissa distance along Cartesian axes and real component of a complex number expressed in Cartesian form (a real quantity)
y	Linear displacement
y	Ordinate distance along Cartesian axes and imaginary component of a complex number expressed in Cartesian form (a real quantity)
$y(t)$	Time domain output of a system excited by a sinusoidal input, Chapter 6
z	Fluid element vertical position above some datum

UPPER CASE LATIN

A	Area
A_{dB}	Amplitude ratio (decibels), Chapter 6
A_j	Numerator coefficient obtained in the partial fraction expansion of N(s)/D(s), Chapter 5
A_1	First denominator factor associated with the transfer function partial fractions in frequency response analysis, Chapter 6
B	Possible amplitude of exciting sinusoid, Chapter 6
C	Fluid capacitance, Chapter 14
C	Leakage coefficient, Chapter 8
C	Temporary working constant, Chapter 6
C_d	Orifice discharge coefficient

Continued on next page

Continued from previous page

C_d	Torque drag coefficient, Chapter 8
C_p	Specific heat of a gas at constant pressure
C_V	Specific heat of a gas at constant volume
C_1, C_2	Numerator coefficients obtained from the partial fraction expansion of $Y(s)$ (complex quantities), Chapter 6
D_m	Motor displacement per radian
D_p	Pump displacement per radian
$D(s)$	Denominator polynomial in the Laplace domain
E	Elastic modulus (linear), Chapter 3
E	Electrical or fluid potential (pressure), Chapter 14
F	Force
$F(s)$	Laplace transform of f(t)
$G(s)$	A general function in the Laplace domain, Chapter 5
$G(s)$	Plant transfer function in the Laplace domain, Chapter 6
$H(s)$	Feedback transducer transfer function in the Laplace domain, Chapter 6
I	Electrical or fluid current (flow rate)
J	Mechanical equivalent of heat
K	Temporary working constant
K_c	Flow gain coefficient for a spool valve
K_h	Fitting head loss factor
K_q	Flow pressure coefficient for a spool valve
K_p	Pressure sensitivity coefficient for spool valve
K_P	Proportional gain for a feedback system
L	Hydraulic inductance
$L(s)$	A general function in the Laplace domain
N	Gearbox ratio
$N(s)$	Numerator polynomial in the Laplace domain
P	Power (mechanical and electrical)

Continued on next page

Continued from previous page

Q	Volumetric flow rate
Q_h	Heat
R	Electrical or hydraulic resistance
Re	Reynolds number
S	Wetted perimeter
T	Torque
$T(s)$	System transfer function in the Laplace domain
$U(s)$	Generic input to a system in the Laplace domain
V	Volume
$V(t)$	Applied fluid pressure function
W	Mechanical work
X	Cartesian coordinate direction
$X(s)$	Laplace transform of linear displacement, x(t)
Y	Cartesian coordinate direction
$Y(s)$	Laplace domain of system excited with sinusoidal input
Z	Cartesian coordinate direction

LOWER CASE GREEK

α	Swash plate displacement angle
β	Bulk modulus
γ	Ratio of the specific heat of gases, Chapter 2
γ	Shear strain, Chapter 2
δ	Various angles relating to an axial piston pump porting geometry
ζ	Damping coefficient (dimensionless)
η	Efficiency
θ	Angle (radians)
μ	Coefficient of absolute viscosity
μ_{st}	Coefficient of static friction
ν	Coefficient of dynamic viscosity

Continued on next page

Continued from previous page

π	Mathematical constant (3.1416)
ρ	Density
τ	Shear stress, Chapter 2
τ	Time constant, Chapter 5
ϕ	Phase angle
ω	Frequency or speed (radians per second)

UPPER CASE GREEK

Θ	Temperature

MISCELLANEOUS

$\mathscr{L}()$	Laplace transform operation

1

INTRODUCTION

1.1 WHAT IS FLUID POWER?

Utilization of fluid power is important because it is one of the three available means of transmitting power. Other methods of transmitting power are by utilizing mechanical means and by applying electrical energy. To demonstrate this we will consider that we have a prime mover such as a diesel engine on one side of the room and a mechanical contrivance on the other. The objective is to see how, in a generic sense, power can be used by the methods quoted above to perform the necessary mechanical work.

For mechanical power transmission, the prime mover is connected to the device and, by use of gearboxes, pulleys, belts and clutches, the necessary work can be performed. With the electrical method, an electrical generator is used. The current developed can be carried through electrical cable to operate electrical motors, linear or rotary, modulation being provided by variable resistance or solid state devices in the circuits. For fluid power utilization, an oil pump is connected to the engine and instead of electrical cables, high pressure hose is used to convey pressurized fluid to motors (again linear or rotary), pressure and flow modulation now being provided within the motors or by means of hydraulic valves. Any of the three methods described may be used however, if an engineering system requires:

1. Minimum weight and volume

2. Large forces and low speeds

3. Instant reversibility

1

4. Remote control

then the fluid power technique will often have significant competitive advantages.

It is indeed unfortunate that design of fluid power systems is seldom taught at four-year universities in the United States at the same time as formal teaching of power transmission systems involving mechanical and electrical systems. Such comprehensive design teaching would demonstrate adequately the advantages of such systems. In some instances hydraulics power transmission is the only technique that can be used. The most spectacular example is that of extending an aircraft's control surface into a high velocity airstream where the only technique available is that of using fluid power actuators because of their high power to weight and volume to weight advantages.

1.2   A BRIEF HISTORY OF FLUID POWER

The performance of mechanical work using pressurized and moving fluids dates back for nearly six millennia. The Egyptians and Chinese used moving water and wind to do work and records show that the advanced civilization in China in 4000 B.C. constructed and utilized wooden valves to control water flow through pipes made of bamboo. In Egypt, the Nile River was dammed so that irrigation could be performed. The Roman Empire also used aqueducts, reservoirs and valves to carry water to cities.

The above applications did of course use dynamic properties of fluids and kinetic energy was employed to perform useful work. Fluid power is, however, customarily associated with the use of potential energy in pressurized fluids. The nearest example in antiquity which comes to mind is the quarrying of marble where holes were drilled in its surface, the holes were then filled with water and the water compressed by hammering in wooden plugs. Pressures achieved as a result were sufficiently high to fracture the marble.

Little scientific progress was made in the Middle Ages in connection with fluid power and it was not until 1648 that a Frenchman, Blaise Pascal, formulated the law that states that pressure in a fluid is transmitted equally in all directions. Practical use was made of this theory by the Englishman Joseph Bramah who built the first hydraulic press in the year 1795. Approximately 50 years later, the Industrial Revolution in Great Britain led to further development of the water press and other industrial machines. The growth was so rapid that by the late 1860s large cities had central fluid power generating stations from which pressurized fluid was pumped to factories. The development of internal combustion engines, manual and automatic controls, and electrical power during the latter part of the nineteenth century, however, diminished the rate of growth of centralized fluid power plants and the practice of such activity ceased.

Interest returned to fluid power at the century's end due to its recognized unique advantages, and in 1906 the electric system for elevating and training guns in the battleship U.S.S. Virginia was replaced by a hydraulic system. In this installation a variable speed hydrostatic transmission system was used to maneuver the guns. Modern ships now make extensive use of fluid power for many services including winches, controllable pitch propellers, rudder control, heavy freight elevators, and raising ammunition from magazines to the guns.

The whole science of fluid power is concerned with the utilization of either a liquid or a gas as a fluid medium. Water and air were the media first used. For some time, however, hydrocarbon based fluids (i.e. oils) have been the dominant liquids. Water based liquids are still used for specialized applications where the flammability of hydrocarbon fluids is unacceptable. In this text, we will be dealing exclusively with hydrocarbon fluids and thus knowledge of their characteristics is required. It should be noted, however, that there exists a fully developed comprehensive technology centered around pneumatic systems and there is significant industrial information and manufacturing activity. The reader is referred to other sources for this information.

1.3   FLUID POWER APPLICATIONS, PRESENT AND FUTURE

Current activity in fluid power technology includes its use to perform transmission and control functions. The growing field of robotics is giving the engineer the opportunity to perform sophisticated design studies for equipment used in many productive sectors such as aerospace, agriculture, automated manufacture, construction, defense, energy and transportation. The above gives an indication of present and future career opportunities for those with skills and experience in fluid power technology. With their increasing use, it is predicted that fluid power components will become less expensive, thereby further improving the competitive advantages of utilizing fluid power as a power transmission medium.

With regard to fluid power components, considerable improvements have been made in the design of seals, fluids, valves, conductors, pumps and motors. The most significant advances in hydraulic system design, however, are seen in the area of controls. Electro-mechanical controls have diversified considerably and have led to many new hydraulic applications.

More recent developments have included the use of programmable controllers in conjunction with hydraulic systems. These controllers contain digitally operated electronic components and have programmable memory with instructions to implement functions such as logic, sequencing, timing, and counting. Such modules may control many different types of machines or processes. It is pleasing to note that fluid power applications are being extended and should increasingly improve our quality of life by, among other things, reducing the need for manual work to be performed.

Dependability has been improved by the development of easily serviced cartridge-type control valves with very long service life and minimum maintenance. Due in part to greater demand, the above systems have been reduced in cost, high pressure piping has been minimized, performance has been improved, and there has been a simplification of maintenance procedures.

As will be demonstrated in this text, the improvements in physical equipment have been accompanied by an enhanced ability to analyze the performance of fluid power systems. Much of this development can be attributed to the dramatic improvement in computing power available to the engineer. More comprehensive analysis will provide a new level of performance for power transmission systems for machines of today and for the future.

1.4   ADVANTAGES OF USING FLUID POWER SYSTEMS

It was stated earlier that there are advantages to using hydraulic systems rather than mechanical or electrical systems for specific applications and for those applications using large powers. Some of these advantages are given below:

1. Force multiplication is possible by increasing actuator area or working pressure. In addition, torques and forces generated by actuators are limited only by pressure and as a result high power to weight ratio and high power to volume ratio are readily achievable.

2. It is possible to have a quick acting system with large (constant) forces operating at low speeds and with virtually instant reversibility. In addition, a wide speed range of operating conditions may be achieved.

3. A hydraulic system is relatively simple to construct with fewer moving parts than in comparable mechanical or electrical machines.

4. Power transmission to remote locations is also possible provided that conductors and actuators can be installed at these locations.

5. In most cases the hydraulic fluid circulated will act as a lubricant and will also carry away the heat generated by the system.

6. A complex system may be constructed to perform a sequence of operations by means of mechanical devices such as cams, or electrical devices such as solenoids, limit switches, or programmable electronic controls.

1.5   A PROBABLE FUTURE DEVELOPMENT

An example of a future development is the design, construction and marketing of a hybrid vehicle, where, instead of using electric generator/motor and power electronics, a hydraulic hybrid design could be advantageous. The U.S. Environmental Protection Agency has already built such a device that has achieved a fuel consumption saving of 55%. The conventional drive train from a stock 2003 four-wheel drive Ford Expedition was removed and replaced by a hydraulic drive train [1]. In addition, the 5.4 L V8 gasoline engine was replaced by a 1.9 L Volkswagen 4 cylinder diesel engine. The hydraulic system used two pumps and two accumulators. One of the pump motors switched between pumping and driving modes and pre-charged the accumulators. During braking, the other pump motor helped to recover braking energy. The pump motor units worked together to pressurize one accumulator to 5000 $lb_f/in.^2$ and the other to 200 $lb_f/in.^2$. It is conjectured that hydraulic hybrid drive trains are particularly well suited to be used in frequently stopping vehicles such as school buses and urban delivery trucks because the system captures large amounts of energy normally lost in braking in conventionally powered vehicles.

REFERENCES

1. ASME, 2004, Mechanical Engineering, **126**(9), p. 13.

2

PROPERTIES OF FLUIDS AND THEIR UNITS

2.1 BASIC PROPERTIES OF FLUIDS

A fluid is defined as a substance that cannot sustain a shearing stress. A fluid can be liquid or gaseous. The science of fluid power is concerned with the utilization of pressurized liquid or gas to transmit power, but we will be dealing exclusively with hydraulic fluids (i.e., liquids). Many textbooks have units of measurement in the U.S. Customary system based upon the former British (or Imperial) system. The use of the more recently defined S.I. system is becoming more common in U.S. industry and for this reason practicing engineers will have to be familiar with both U.S. basic and S.I. systems of units (see Table 2.1).   In many fluid power applications the inch is used as the basic unit of length. Later in this section the use of these units will be demonstrated (see Table 2.2) and some of the special problems that may occur with the use of mass in the U.S. Customary form of units will be discussed.   Knowledge of the individual characteristics of hydraulic fluids is essential and this section deals with their fundamental properties.

Oil density: This is defined as mass per unit volume. For petroleum based hydraulic fluids the approximate value is $\rho = 850 \text{ kg/m}^3$. It should be observed that a dynamic analysis that uses $\text{lb}_f/\text{in.}^2$ as a pressure unit must be consistent and use mass in $\text{lb}_f \cdot \text{s}^2/\text{in}$. Accelerations will be in $\text{in.}/\text{s}^2$. Unfortunately there is no special name for a mass unit in the pound force, inch, second system. A mass unit in the pound force, foot, second system

7

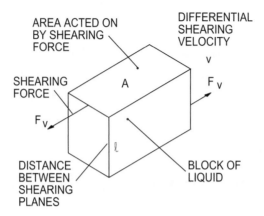

Figure 2.1: Definition of absolute viscosity.

is the *slug* where 1 slug = 1 lb$_f$ · s^2/ft. Some authors use weight per unit volume, then the term *Specific Weight* should be used.

Specific gravity: This is the ratio of the mass of a substance divided by the mass of an equal volume of water at some specified temperature, usually 20°C. The unit is therefore dimensionless and varies between 0.8 for some petroleum based fluids to as high as 1.5 for the chlorinated hydrocarbons.

Absolute viscosity: This is a measure of the resistance to motion offered by a fluid caused by the generation of shear stress over a wetted area (Figure 2.1). The resistance is therefore proportional to the wetted area and to the velocity and inversely proportional to the film thickness.

$$\frac{F_v}{A} = \mu \frac{v}{\ell}$$

or:

$$\mu = \frac{F_v}{A(v/\ell)} = \frac{\tau}{\dot{\gamma}} \tag{2.1}$$

For a piston concentrically located in a circular cylinder with oil in the clearance space, ℓ_1, the area A is given by $A = \pi d\ell_2$ and the force F_v is given by:

$$F_v = \frac{\pi d\ell_2 \mu v}{\ell_1} \tag{2.2}$$

Table 2.1: Conversion between U.S. Customary and SI units

Basic US Customary to S.I.	Unit	S.I. to Basic U.S. Customary
1 in. = 25.40 mm	Length	1 m =39.37 in.
1 ft = 0.3048 m		1 m = 3.281 ft
1 lb$_f$ = 4.448 N	Force	1 N = 0.2248 lb$_f$
1 slug = 14.594 kg	Mass	1 kg = 0.0685 slug
1 lb$_f$ · s^2/in. = 175.128 kg		1 kg = 0.00571 lb$_f$ · s^2/in.
1 slug/ft^3 = 515.4 kg/m^3	Density	1 kg/m^3 = 1.94E−3 slug/ft^3
1 lb$_f$ · s^2/in^4. = 10.69E+6 kg/m3		1 kg/m^3 = 93.57E−9 lb$_f$ · s^2/in^4.
1 K = 5(°F − 32)/9	Temperature	°F = 32 + 1.8(K − 273.2)
°R = °F + 460		K = °C + 273.2
1 lb$_f$ · s/in.^{2} = 1 reyn (= 68.97E+4 Poise)	Absolute Viscosity	1 MPa · s = 145 lb$_f$ · s/in.^{2} (= 10E+7 Poise)
1 psi = 6.985 kPa	Pressure	1 MPa = 145.0 lb$_f$/in.^{2}
1 ksi = 6.985 Mpa	(or Stress)	1 MN/m^2 = 145.0 psi
1 lb$_f$ft = 1.356 N · m	Torque	1 N · m = 0.7376 lb$_f$ · ft
1 ft · lb$_f$ = 1.356 J	Work or	1 J = 0.7376 ft · lb$_f$
1 Btu = 1054 J	Energy	1 J = 0.968E−3 Btu
1 hp = 745.7 W	Power	1 kW = 1.341 hp

The quantity μ is termed the coefficient of absolute viscosity. Conversion between various sets of units can be confusing. Noting that absolute viscosity has the fundamental dimensions F.T/L^2, which is equivalent to M/L.T, will allow conversions to be made on a rational basis. Often we measure absolute viscosity in lb$_f$ · s/in.^{2} units (called the reyn). Sometimes units using centimeter, gram and second are used (the cgs system), then the unit is called the Poise for viscosity. Thus 1 centipoise = 1 cP = 1.0E−2 Poise = 1 mPa · s = 1.45E−7 reyn.

The relationship between absolute viscosity and temperature is very non linear. This is shown in Figure 2.2 where the abscissa values are absolute temperature values divided by a room temperature of 68°F (20°C) also converted to absolute. The ordinate values are the absolute viscosity normalized with respect to the absolute viscosity at room temperature. Probably the most important feature of Figure 2.2 is that it shows that fluid power actuators and controls on offroad equipment initially operating at very low temperatures will operate erratically or very slowly until the oil temperature has risen to a higher value. The graph also shows that calculations that assume turbulent flow and viscosity independence are likely to be quite accurate at normal operating temperatures between 68 and 95°F (20 and 35°C).

Oil refiners use the term *viscosity index* to describe the degree of the change in viscosity with temperature. The viscosity index is discussed in more depth in [1].

Kinematic viscosity: This is the ratio of viscosity to the mass density. Thus:

$$\nu = \frac{\mu}{\rho} \qquad (2.3)$$

In the cgs system the unit is the Stoke, but the centistoke (1/100 of a Stoke) is a more convenient size. Kinematic viscosity is difficult to measure directly, so an indirect (empirical) measurement is used. The flow of a known quantity of liquid through a given sized orifice under gravity is timed.

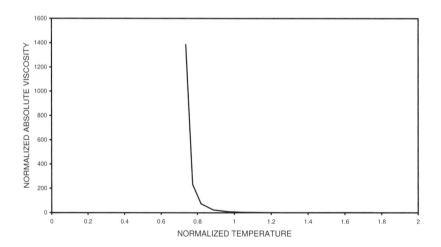

Figure 2.2: Change of absolute viscosity with temperature, 1 is room temperature.

Because of the known variation of viscosity with temperature, the apparatus is usually contained in a water bath so this characteristic can be controlled. The best known unit in the U.S. is the Saybolt second (SSU) and we may write [2]:

$$\nu = 0.216 \text{ SSU} - \frac{166}{\text{SSU}} \text{ centistokes (cS)} \qquad (2.4)$$

Because density changes much less rapidly with temperature than does absolute viscosity, the relation between kinematic viscosity and temperature closely resembles that shown in Figure 2.2.

The American Society for Testing Materials has determined that certain logarithmic transformations will allow most common hydrocarbon fluids to be displayed as straight lines on specialized kinematic viscosity vs. temperature charts over temperature ranges that will be met in practice. Such information is presented in ASTM Standard D341 [3].

2.1.1 Example: Conversion Between Viscosity Units

Consider a petroleum based fluid with a viscosity of 14.3 cS and a density of 0.84 g/cm^3 at 38°C. Determine the absolute viscosity of this oil in inch pound force units. The centistoke is a cgs unit so we can write:

$$\nu = 14.3 \text{ cS} = 143\text{E}{-}3 \ \frac{\text{cm}^2}{\text{s}}$$

The kinematic viscosity in inch units is given by:

$$
\begin{aligned}
\nu &= 143\text{E}{-}3 \frac{((1/2.54)^2 \text{ in.}^2/\text{cm}^2)\text{cm}^2}{\text{s}} \\
&= 22.17\text{E}{-}3 \ \frac{\text{in.}^2}{\text{s}}
\end{aligned}
$$

We now need to find the density in inch units. Start with the standard mass unit the pound mass. This has the conversion between pound mass and the gram of:

$$1 \text{ lb}_\text{m} = 453.59 \text{ g}$$

so the first step in calculating density is:

$$
\begin{aligned}
\rho &= 0.84 \frac{((1/453.59) \text{ lb}_\text{m}/\text{g})\text{g}}{((1/2.54)^3 \text{ in.}^3/\text{cm}^3)\text{cm}^3} \\
&= 0.03041 \ \frac{\text{lb}_\text{m}}{\text{in.}^3}
\end{aligned}
$$

Next we need to convert from pound mass to the mass unit consistent with a force in lb_f and an acceleration of 1 in./s^2. This conversion is:

$$1 \frac{lb_f \cdot s^2}{in.} = 1 \times 32.174 \times 12 = 386.09 \ lb_m$$

Hence density in compatible inch units is:

$$
\begin{aligned}
\rho &= 0.03041 \frac{lb_m}{in.^3} \times \frac{1 \ lb_f \cdot s^2/in.}{386.088 \ lb_m} \\
&= 78.76E{-}6 \frac{lb_f \cdot s^2}{in.^4}
\end{aligned}
$$

The absolute viscosity may now be calculated in U.S. Customary inch units:

$$
\begin{aligned}
\mu = \rho\nu &= 78.76E{-}6 \frac{lb_f \cdot s^2}{in.^4} \times 2.217E{-}02 \frac{in.^2}{s} \\
&= 1.746E{-}06 \frac{lb_f \cdot s}{in.^2} = 1.746 \ \mu\text{reyn}
\end{aligned}
$$

Specific heat: The specific heat of oil is approximately 0.5 Btu/lb·° F (in SI units 2.1 J/g · K).

Thermal conductivity: The thermal conductivity of oil is approximately 0.08 (Btu/h · ft^2)/(°F/ft) (in SI units 0.14 W/m · K).

2.2 COMPRESSIBILITY OF LIQUIDS

In most fluid mechanics classes that are taught to undergraduates, liquids are treated as incompressible. A fluid power system using a liquid such as a hydrocarbon liquid operates at pressures where the compressibility of a liquid has a noticeable effect on the operation of a system. It will be seen in Chapter 4, that one of the fundamental modeling equations involves dp/dt. In turn this rate of change of pressure equation involves compliance, i.e. the effect of liquid compressibility. Bulk modulus is an elastic constant giving the amount that the oil volume is reduced for a given application of pressure. This property of compressibility of oil is an important property because:

1. It has a large influence upon the fundamental frequency of a fluid power system, which will be shown later to have a significant effect upon the output speed of response.

2. The fact that a differential coefficient is involved in its definition makes it possible to determine the change of pressure in a fluid power system.

When a change in pressure is applied to an oil volume, a volume change occurs. The relation between these quantities can be written:

$$\beta = -\frac{\Delta p}{\Delta V / V} \quad (2.5)$$

This definition is the ratio stress/strain and in this case applied pressure may be regarded as the stress, and dV/V is the resulting volumetric strain. In addition, since the volume is reduced for a positive pressure application (i.e. $\Delta V < 0$), a negative sign is necessary in the expression for bulk modulus.

Effective bulk modulus: The effective bulk modulus has to be used for fluid power systems. This is because the containing vessels and undissolved air may significantly reduce the system bulk modulus [2, 4]. Consider the configuration shown in Figure 2.3. If the rigid piston is displaced ΔV, the pressure in the system will rise Δp. The decrease in system volume is the sum of the compression of the oil and of the air. In formal terms:

$$\frac{1}{\beta_e} = -\frac{\Delta V}{\Delta p V} = -\frac{\Delta V_{OIL} + \Delta V_{AIR}}{\Delta p V_T}$$

Rearranging slightly:

$$\frac{1}{\beta_e} = \frac{-(\Delta V_{OIL}/V_{OIL})}{\Delta p}\frac{V_{OIL}}{V_T} + \frac{-(\Delta V_{AIR}/V_{AIR})}{\Delta p}\frac{V_{AIR}}{V_T}$$

Thus the exact expression is:

$$\frac{1}{\beta_e} = \frac{1}{\beta_{OIL}}\left(\frac{V_T - V_{AIR}}{V_T}\right) + \frac{1}{\beta_{AIR}}\frac{V_{AIR}}{V_T} \quad (2.6)$$

Because $V_{AIR}/V_T \ll 1$ a simplified estimate for the effective bulk modulus is commonly used:

$$\frac{1}{\beta_e} = \frac{1}{\beta_{OIL}} + \frac{V_{AIR}}{V_T}\frac{1}{\beta_{AIR}} \quad (2.7)$$

It should be noted that the reciprocal nature of Equation 2.7 means that $\beta_e < \beta_{OIL}$.

Processes that occur during the operation of fluid power devices occur sufficiently rapidly that they may generally be considered adiabatic. Thus the correct relation between pressure and volume for air is:

$$p = \frac{K_{AIR}}{V_{AIR}^{\gamma_{AIR}}} \quad (2.8)$$

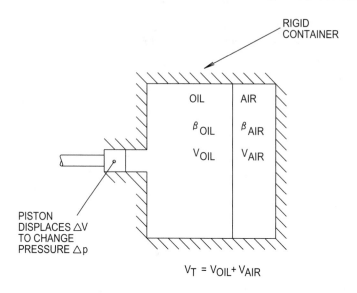

Figure 2.3: Bulk modulus of an oil and undissolved air mixture.

Differentiating this expression and using the basic definition of bulk modulus shows that the bulk modulus of air is:

$$\beta_{AIR} = \gamma_{AIR}\, p \qquad (2.9)$$

A numerical comparison may be useful. A typical hydrocarbon oil has a bulk modulus of 1.86E+3 MPa (266E+3 lb$_f$/in.^{2}) when devoid of entrained air. If the system is operating at 7 MPa, then the undissolved air will have a bulk modulus of only 9.8 MPa (noting $\gamma_{AIR} = 1.4$). Consequently only a small amount of entrained air will lower the effective bulk modulus significantly.

Although the affect of undissolved air is to reduce effective bulk modulus significantly, the effect is reduced at higher working pressures [4]. Consider a volume of oil and undissolved air that is subject to a change in pressure from 0 to 20 MPa gauge. It will be assumed that the oil remains at constant temperature and the air does not dissolve as the pressure rises. The volume of the oil component will be given approximately by:

$$V_{OIL} = V_{OIL0}(1 - p/\beta_{OIL}) \qquad (2.10)$$

In fact, because working pressures are much less than the magnitude of the bulk modulus, the oil volume may be considered unchanged for the purpose of this calculation. On the other hand, the undissolved air volume will change appreciably:

$$V_{AIR} = \frac{p_0 + p_{atm}}{p + p_{atm}} V_{AIR0} \qquad (2.11)$$

If these two expressions are used to calculate the effective bulk modulus using Equation 2.6, then Figure 2.4 is obtained. Several conclusions may be drawn from the figure. First a well designed system that brings oil into reservoirs at atmospheric pressure must be designed to limit oil entrainment. Assume that 5% air is entrained at atmospheric pressure. If the system pressure is 7 MPa (1000 lb$_f$/in.^{2}) or more, which is a reasonable requirement, then the effective bulk modulus will be at least 80% of the uncontaminated oil value. The second feature of Figure 2.4 is the asymptotic approach of effective bulk modulus to the uncontaminated the bulk modulus at high pressures. Consequently any system used for precise position or velocity control ought to operate at as high a pressure as is appropriate or practical.

Effect of hoses: Unfortunately other factors can contribute to a reduced effective bulk modulus. Probably the major components that affect

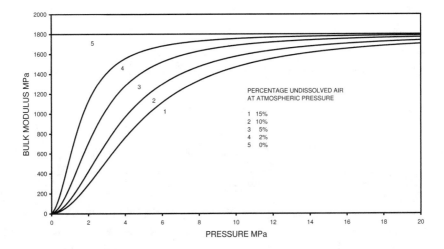

Figure 2.4: Effect of undissolved air on effective bulk modulus.

bulk modulus adversely are flexible hoses. These are constructed of a polymer adjacent to the oil, one or more layers of steel braid, and a polymer
cover. Although the polymer layer adjacent to the oil is kept as thin as
practical, it will have a modulus much less than steel. Also the steel braid
is not as rigid as a steel pipe of the same diameter and wall thickness. The
effect of one or more volumes that have a bulk modulus significantly less
than the oil in them can be estimated with the assistance of Figure 2.5. As
with the previous derivation for oil and air, we consider an increase in pressure on the system of Δp. In this development, we shall use the subscript
O in place of OIL to denote the oil bulk modulus and volume. This is to
indicate that the oil may be contaminated with undissolved air. There will
be an associated change in system volume. This volume is comprised of the
reduction in volume of the oil and an increase in the container volumes:

$$\Delta V = \Delta V_O + \sum_{r=1}^{r=n} \Delta V_r$$

As before, we write the effective bulk modulus of the oil in the system as:

$$\beta_e = -\frac{\Delta p}{\Delta V / V_T}$$

This expression is inverted to yield:

$$\frac{1}{\beta_e} = -\frac{\Delta V_O + \sum_{r=1}^{r=n} \Delta V_r}{\Delta p V_T}$$

Modify this to obtain the form:

$$\frac{1}{\beta_e} = -\left(\frac{\Delta V_O}{\Delta p V_T} + \frac{\sum_{r=1}^{r=n} \frac{\Delta V_r V_r}{V_r}}{\Delta p V_T} \right)$$

This can be written in the more comprehensible form:

$$\frac{1}{\beta_e} = \frac{1}{\beta_O} + \sum_{r=1}^{r=n} \frac{f_r}{\beta_r} \tag{2.12}$$

Where the f_r quantities represent the fraction of the total oil volume
contributed to by container r.

Estimates of container bulk modulus for thick walled cylinders may be
found in strength of materials texts. Fortunately thick walled vessels are
very rigid and do not contribute significantly to the reduction in effective
bulk modulus. For a thick walled cylinder, approximately:

$$\beta_C = \frac{E}{2.5} \tag{2.13}$$

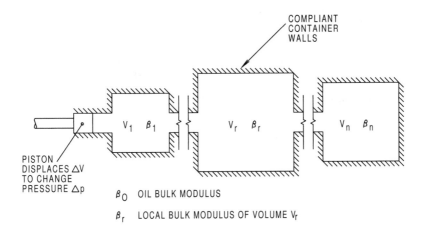

Figure 2.5: Bulk modulus of multiple compliant volumes in a system.

For a thin walled cylinder, of wall thickness ℓ_w and internal diameter d, we may write:

$$\beta_C = \frac{\ell_w E}{d} \qquad (2.14)$$

2.2.1 Example: Bulk Modulus of Multiple Containers

Consider a system consisting of a pump, a valve, a flexible hose, and a cylinder. The salient properties of the system are shown in Table 2.2. Calculate the system bulk modulus accounting for entrained air and the compliance of the containing volumes.

Calculate the volumes of the hose and the cylinder:

$$\text{Hose volume} = \frac{\pi 0.5^2}{4} \times 60 = 11.78 \text{ in.}^3$$

$$\text{Cylinder volume} = \frac{\pi 2^2}{4} \times 36 = 113.1 \text{ in.}^3$$

$$\text{Total fluid volume} = 10 + 11.38 + 113.1 = 134.9 \text{ in.}^3$$

Table 2.2: Properties of a system using different container characteristics

Characteristic	Size	Units
Pump and valve volume	10.0	in.^{3}
Pump wall material linear modulus	30E+6	lb$_f$/in.^{2}
Flexible hose length	60.0	in.
Flexible hose inner diameter	0.5	in.
Flexible hose bulk modulus [4]	7000	lb$_f$/in.^{2}
Cylinder length	36.0	in.
Cylinder inner diameter	2.0	in.
Cylinder wall thickness	0.075	in.
Cylinder wall material linear modulus	30E+6	lb$_f$/in.^{2}
Uncontaminated oil bulk modulus	0.26E+6	lb$_f$/in.^{2}
Undissolved air volume in reservoir at atmospheric pressure	5.0	%
System operating pressure	1000	lb$_f$/in.^{2} gauge
Atmospheric pressure	14.7	lb$_f$/in.^{2} abs.

Calculate bulk modulus for each of the compliant container volumes:

$$\text{Pump bulk modulus, Equation 2.13} = \frac{30E+6}{2.5} = 12E+6 \text{ lb}_f/\text{in.}^2$$

$$\begin{aligned} \text{Cylinder bulk modulus, Equation 2.14} &= \frac{0.075 \times 30E+6}{2} \\ &= 1.125E+6 \text{ lb}_f/\text{in.}^2 \end{aligned}$$

The oil with the undissolved air must also have its bulk modulus evaluated at a working pressure of 1000 lb$_f$/in.^{2} The following calculation assumes that the temperature of the oil in the reservoir is equal to the temperature in the system. Although this is unlikely to be exactly true, the approximation will be adequate in most instances:

$$\text{Oil volume} = V_{O0}\left(1 - \frac{p}{\beta_O}\right) = 0.95\left(1 - \frac{1000}{260000}\right)$$
$$= 0.9463$$

$$\text{Air volume} = V_{A0}\left(\frac{p_{atm}}{p_{atm} + p}\right) = 0.05\left(\frac{14.7}{14.7 + 1000}\right)$$
$$= 0.0007244$$

$$\text{Total volume} = 0.9463 + 0.0007244$$
$$= 0.947$$

Use Equation 2.9 to calculate the bulk modulus of air when compressed to 1000 $\text{lb}_\text{f}/\text{in.}^2$:

$$\beta_A = \gamma_A p = 1.4\,(14.7 + 1000)$$
$$= 1421 \ \text{lb}_\text{f}/\text{in.}^2$$

Noting that the pressure in this expression is an absolute pressure.

Now use Equation 2.6 to calculate the effective bulk modulus of the oil contaminated with air:

$$\frac{1}{\beta_{OIL}} = \left(\frac{V_T - V_A}{V_T}\right) + \frac{1}{\beta_A}\frac{V_A}{V_T}$$
$$= \frac{1}{0.26\text{E}+6}\frac{0.9463}{0.947} + \frac{1}{1421}\frac{0.0007244}{0.947}$$
$$= 4.382\text{E}{-}6$$

Although we shall only need to use the reciprocal of the bulk modulus, we will calculate the bulk modulus of the liquid alone under working conditions:

$$\beta_{OIL} = 0.2282\text{E}+6 \ \text{lb}_\text{f}/\text{in.}^2$$

We shall now use Equation 2.12 to calculate the effective bulk modulus of the oil in the system accounting for the compliancies of the various containing volumes and the air in the oil:

$$\frac{1}{\beta_e} = \frac{1}{\beta_{OIL}} + \sum_{r=1}^{r=n} \frac{f_r}{\beta_r}$$

$$= \frac{1}{0.2282E+6} + \frac{10}{134.9} \times \frac{1}{12E+6}$$

$$+ \frac{11.78}{134.9} \times \frac{1}{7E+3}$$

$$+ \frac{113.1}{134.9} \times \frac{1}{1.125E+6}$$

$$= 4.382E-6 + 0.006177E-6$$

$$+ 12.47E-6 + 0.745E-6$$

$$= 17.6E-6$$

So the effective bulk modulus of the oil in the system is:

$$\beta_e = 56.8E+3 \text{ lb}_f/\text{in.}^2$$

Before leaving this example, it should be noted that calculations of this nature can seldom be performed with such apparent accuracy because the input information will not be known very accurately. The example should be used to gain some appreciation of the relative magnitude of compliance effects. Inspect the terms in the specific application of Equation 2.12 to find the terms with the largest values. Because the equation has the bulk moduli in reciprocal form, it is these relatively large terms that will dominate the final inversion. In this example, the flexible hose term at $12.47E-6$ is easily the largest term and the air contaminated oil term follows at $4.382E-6$. Although a thin walled cylinder was chosen, its effect is less than one tenth of the hose term. In summary, using flexible hoses has been the main factor in reducing the bulk modulus of the uncontaminated oil from $260E+3$ to a system value of $56.8E+3$ lb$_f$/in.^{2} For many purposes, such a reduction may be of no significance. On the other hand, any system where precise positional or velocity control is desired will need attention to maintaining high values of effective bulk modulus for the system.

In summary, bulk modulus of a typical hydrocarbon based fluid with no entrapped air is about $270,000$ lb$_f$/in.^{2} (or 1860 MPa). Entrapped air, however, has a significant influence upon its value and typically the effective value in a working system may be $200,000$ lb$_f$/in.^{2} (or 1380 MPa) at best and often much less if flexible hoses are present.

The oil spring: Another effect of bulk modulus is that any constrained oil volume behaves like a spring. Figure 2.6 shows a double acting cylinder.

There is oil on both sides of the piston. For generality, the two sides of the cylinder are not matched for piston area. Consider the situation where the oil flow to or from the cylinder is blocked so volumes V_1 and V_2 are trapped. The effective bulk modulus of the trapped oil accounting for entrapped air and hose and cylinder compliance is β_e. Suppose a incremental force ΔF is applied to the piston and the rod moves Δx relative to the cylinder. The pressure in V_1 will fall by Δp_1 and the pressure in V_2 will rise by Δp_2. If this were a conventional metallic spring, we would write the expression for stiffness as:

$$k = \frac{\Delta F}{\Delta x} \tag{2.15}$$

Using the characteristics of the cylinder, this expression can be written:

$$k = \frac{-\Delta p_1 A_1}{\Delta V_1 / A_1} + \frac{\Delta p_2 A_2}{-\Delta V_2 / A_2}$$

The sign convention implies that an increase in pressure is positive and that an increase in fluid volume is likewise positive. Now recast Equation 2.5 in the form:

$$-\frac{\Delta p}{\Delta V} = \frac{\beta}{V}$$

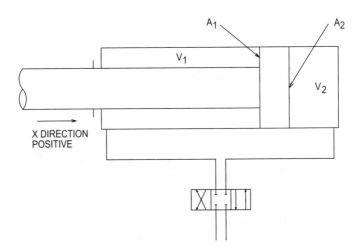

Figure 2.6: Trapped oil in a cylinder treated as a spring.

Hence the spring rate expression can be written:

$$k = \frac{\beta_e A_1^2}{V_1} + \frac{\beta_e A_2^2}{V_2}$$
$$= \beta_e \left(\frac{A_1^2}{V_1} + \frac{A_2^2}{V_2} \right) \tag{2.16}$$

We now need to investigate the variation of stiffness as the relative values of V_1 and V_2 change. This is most easily done by introducing a total oil volume V_T that does not change as the piston moves. The stiffness expression will then have the form:

$$k = \beta_e \left(\frac{A_1^2}{V_1} + \frac{A_2^2}{V_T - V_1} \right)$$

This expression may be plotted or the rules of calculus applied to show that the stiffness has a minimum value when:

$$\frac{A_1^2}{V_1} = \frac{A_2^2}{V_T - V_1} = \frac{A_2^2}{V_2}$$

Now most cylinders that will be used for precise positioning of loads will use a double, symmetrical rod configuration. In the usual situation where the pipe connections from the valve have equal volumes, the *minimum* stiffness condition occurs when the piston is in the center of its travel.

The significance of the oil spring stiffness will be seen later when dynamic models of systems are discussed.

2.2.2 Example: The Oil Spring

Consider the configuration shown in Figure 2.7. Observe that the length of the supply pipe to side 1 of the cylinder is considerably shorter than that to side 2. If both supply pipes have the same diameter, determine the diameter of the supply pipes so the undamped frequency is not less than 640 Hz. Then determine the undamped frequency if the piston were moved to the center of the cylinder.

Many of the examples in this chapter have been solved using in., $\text{lb}_f \cdot \text{s}^2/\text{in.}$, and s units because so much of the fluid power equipment manufactured in the U.S. is still provided in inch dimensions. For pedagogic purposes, this problem will be solved in SI units. A set of characteristics is provided in U.S. Customary units in Table 2.3 and these have been converted to SI units in Table 2.4.

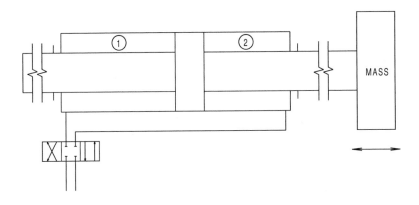

Figure 2.7: Load oscillation caused by oil spring.

The undamped natural frequency of a mass attached to a spring is (for a derivation of the next equation see Section 5.7):

$$\omega_n = \sqrt{\frac{k}{m}} \tag{2.17}$$

In this problem, the cylinder has two symmetrical rods. Thus solving the first part of the problem requires finding the pipe diameter and the piston position so the oil volumes on sides 1 and 2 are equal. Using d_p for the pipe diameter and ℓ_1 for the piston displacement on side 1, then we can write:

$$V_1 = \frac{\pi d_p^2}{4} \times 0.0508 + \frac{\pi(0.0762^2 - 0.0254^2)}{4} \times \ell_1$$

$$V_2 = \frac{\pi d_p^2}{4} \times 1.016 + \frac{\pi(0.0762^2 - 0.0254^2)}{4} \times (0.1016 - \ell_1 - 0.01905)$$

The form of these equations allows easy elimination of ℓ_1:

$$2V_1 = V_1 + V_2 = \frac{\pi d_p^2}{4} \times 1.0668 + \frac{\pi 0.005161}{4} \times 0.08255$$

Table 2.3: U.S. Customary characteristics for the oil spring example

Characteristic	Size	Units
Cylinder internal length	4.0	in.
Cylinder internal diameter	3.0	in.
Rod diameter	1.0	in.
Piston width	0.75	in.
Length of supply pipe to side 1	2.0	in.
Length of supply pipe to side 2	40.0	in.
Mass of load, rod, and piston	23.17	lb_m
Oil bulk modulus	200E+3	$lb_f/in.^2$

Table 2.4: SI characteristics for the oil spring example

Characteristic	Size	Units
Cylinder internal length	0.1016	m
Cylinder internal diameter	0.0762	m
Rod diameter	0.0254	m
Piston width	0.01905	m
Length of supply pipe to side 1	0.0508	m
Length of supply pipe to side 2	1.016	m
Mass of load, rod, and piston	10.51	kg
Oil bulk modulus	1.379E+9	Pa

Thus, for a minimum frequency (i.e., minimum stiffness) condition, the oil volume on each side of the piston is equal to:

$$V_1 = 0.41897d_p^2 + 167.3E{-}6$$

We can now determine the spring stiffness required to achieve a frequency of 640 Hz:

$$k = \omega_n^2 m \ = \ (2\pi \times 640)^2 \times 10.51$$
$$= \ 169.9\text{E}{+}6 \text{ N/m}$$

Using Equation 2.16 for the stiffness of an oil spring and rearranging this to yield the oil volume:

$$k = \beta_e \left(\frac{A_1^2}{V_1} + \frac{A_2^2}{V_2} \right) = 2\beta_e \frac{A_1^2}{V_1}$$

$$V_1 = \frac{2\beta_e A_1^2}{k}$$

Substitute the expressions of volume and stiffness found previously:

$$0.41897 d_p^2 + 167.3\text{E}{-}6 = \frac{2 \times 1.379\text{E}{+}9 \times \left(\frac{\pi}{4}(0.0762^2 - 0.0254^2) \right)^2}{169.9\text{E}{+}6}$$

$$d_p^2 \ = \ \frac{266.7\text{E}{-}6 - 167.3\text{E}{-}6}{0.41897}$$
$$= \ 237.3\text{E}{-}6$$

$$d_p = 15.4\text{E}{-}3 \text{ m} = 0.606 \text{ in.}$$

For the second part of the problem, the undamped frequency is to be calculated when the piston is centered in the cylinder. Here the piston displacement on each side is 0.04123 m. We can use the expressions used previous to calculate the trapped oil volumes:

$$V_1 \ = \ \frac{\pi 0.0154^2}{4} \times 0.0508 + \frac{\pi 0.005161}{4} \times 0.04123$$
$$= \ 176.6\text{E}{-}6 \text{ m}^3$$

$$V_2 \ = \ \frac{\pi 0.0154^2}{4} \times 1.016 + \frac{\pi 0.005161}{4} \times 0.04123$$
$$= \ 356.4\text{E}{-}6 \text{ m}^3$$

For convenience calculate the face area of the piston:

$$A_1 = \frac{\pi(0.0762^2 - 0.0254^2)}{4}$$
$$= 4.054\text{E}{-}3 \text{ m}^2$$

Hence the stiffness is:

$$k = 1.379\text{E}{+}9 \times (4.054\text{E}{-}3)^2 \left(\frac{1}{176.6\text{E}{-}6} + \frac{1}{356.4\text{E}{-}6} \right)$$
$$= 191.9\text{E}{+}6 \text{ N/m}$$

Thus we can calculate the frequency:

$$\omega_n = \sqrt{\frac{191.9\text{E}{+}6}{10.51}} = 4273 \text{ rad/s}$$
$$\equiv 680 \text{ Hz}$$

The oil spring phenomena allows us to comment on some of the compromises that must be made when designing fluid power systems. We have shown that operating at high pressure reduces the effect of undissolved air on bulk modulus. In turn this will raise the stiffness of a system. On the other hand, Equation 2.16 shows that having a large value of piston cross section area will also increase stiffness. These two features may be contradictory. As pressures become higher, the size of equipment becomes smaller. Regardless of pressure, however, Equation 2.16 shows that trapped oil volumes should be minimized to raise stiffness. Any design requiring precise positioning or velocity control should pay close attention to minimizing piping volumes.

PROBLEMS

2.1 A hydrocarbon oil is being used in a circuit connecting a valve to an actuator. The characteristics of the system are presented in the table. The walls of the actuator may be assumed unyielding.

Characteristics of a circuit for determining effective bulk modulus

Characteristic	Size	Units
Length of hose from valve to actuator	2.3	m
Internal hose diameter	15.0	mm
Oil length in actuator cap side	0.5	m
Actuator cap side diameter	100	mm
Air content by volume of oil in reservoir	5.0	%
Pressure on cap side of actuator	10.0	MPa
Bulk modulus of uncontaminated oil	1800.0	MPa
Hose bulk modulus	50.0	MPa
Oil temperature in system	45.0	°C

Find the effective bulk modulus in the volume of oil trapped between the valve and the piston of the actuator.

2.2 Consider the engine hoist shown in the figure. For simplicity, assume the hoist frame and arm are completely rigid. The connections to the actuator are through rigid pipe and the walls of the actuator may be taken as rigid. The characteristics of the system are given in the table.

Find the working pressure when the engine is lifted, assume the arm is horizontal. Find the air fraction in the actuator at working pressure. Find the stiffness of the oil column in the actuator under working conditions. The natural frequency of a mass, m, hung (or supported) on a spring of stiffness, k, is:

$$\omega_n = \sqrt{\frac{k}{m}}$$

You may ignore the mass of the hoist arm. Determine the effective stiffness of the oil spring at the point of suspension of the engine mass and then determine the natural frequency of the suspended engine.

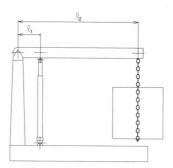

Characteristics of a circuit for determining system natural frequency

Characteristic	Size	Units
Length of hose from valve to actuator	2.3	in.
Length ℓ_1	9.0	in.
Length ℓ_2	48.0	in.
Actuator cap side diameter	1.5	in.
Oil column length	18.0	in.
Engine mass	400	lb_m
Bulk modulus of uncontaminated oil	260E+3	$lb_f/in.^2$
Air content by volume of oil in reservoir	7.5	%
Oil temperature in system	88.0	°F

REFERENCES

1. Wolansky, W. and Akers, A., 1988, *Modern Hydraulics the Basics at Work*, Amalgam Publishing Company, San Diego, CA.

2. Merritt, H. E., 1967, *Hydraulic Control Systems*, John Wiley & Sons, New York, NY.

3. ASTM International, 1998, "Standard Viscosity-Temperature Charts for Liquid Petroleum Products", D341-93, ASTM International, West Conshohocken, PA.

4. McCloy, D. and Martin, H. R., 1980, *Control of Fluid Power: Analysis and Design*, 2nd (revised) edition, Halsted Press: a division of John Wiley & Sons, Chichester, U. K.

3

STEADY STATE MODELING

3.1 RATIONALE FOR MODEL DEVELOPMENT

Mathematical models may be used to develop an understanding of a system and to improve its performance. Models will aid in the selection of components and individual parts that provide desired system behavior. Establishing a mathematical model for a fluid power system is a flexible process. Several different models could be written for a particular system. The type of model used depends on the intended use of the machine and the nature of its parts. A machine may have certain critical performance requirements that influence the approach used in an analytical study. Every characteristic used to describe the machine mathematically will have variations in its numerical value. When machines are put to work they generally must cover variations in work loads and functions. Therefore, the mathematical model must be written to allow easy variation in characteristic values.

In essence, modeling is concerned with the prediction of energy distribution in a system. Energy must be added to the moving fluid stream in a fluid power system in order to produce work at some desired point. Energy can be accounted for at any point in the system at all times as the machine performs its functions. Some of the pressure energy in the system is converted into heat and is not available for useful work. It is common practice to call the conversion of pressure energy into heat a *loss*, but it should be realized that energy is not *lost*, only converted. Losses occur due to fluid friction as the fluid flows through tubing, hoses, fittings, and valves. Friction between moving mechanical parts also causes losses. The system output devices then utilize the remaining pressure energy. Writing a suitable model allows the designer to predict the output power that will be available from the system and the capacity of heat exchangers required

31

to maintain acceptable fluid temperatures.

An energy source, such as an internal combustion engine or electric motor, is needed to drive a fluid power machine. A hydraulic system is never a source of energy, it is in reality an energy transfer system.

3.2   SOURCE OF EQUATIONS

The design and development of hydraulic machinery can be greatly enhanced with the use of appropriate equations. These equations, when arranged in mathematical models, can be used to describe the performance of machines under operating conditions.

In this chapter, the conditions in the circuit will be considered invariant with time, that is to say *steady state*. In two of the examples shown, the flow regulator valve and the accumulator, the conditions are not steady state. Conditions may sometimes be relaxed and steady state equations used when dynamics (e.g., $F = ma$) and fluid compliance (e.g., $dp/dt = (\beta/V)dV/dt$) have only an insignificant effect. In Chapter 4 and in the latter part of the text, dynamic systems where conditions vary with time will be modeled. In essence, steady state modeling can be performed with algebraic equations, but time varying conditions usually require a combination of algebraic and differential equations.

Equations will be drawn from several engineering subjects. When fluid passes through devices and passages, there will be a conversion of pressure (effectively potential) energy to mechanical energy (i.e., work) and to heat energy. Equations from fluid mechanics will be used to calculate pressure drops and equations from thermodynamics will be used to determine energy conversion. These hydraulic power equations may then be combined with appropriate equations that describe power transmission, linkages, and possibly electronic devices to provide a complete machine model.

Any liquid can be used to transmit fluid power. From a practical standpoint, however, most of the fluids used in hydraulic systems are petroleum based oils that are Newtonian in nature [1]. If non-Newtonian liquids are used, such as water-in-oil emulsions, consideration must be given to use of appropriate values for the absolute viscosity, μ.

Any convenient system of units could be used to study fluid power machinery. This text is mainly concerned with the development of mathematical expressions useful for describing the transmission of power in hydraulic systems. Therefore, the unit systems that are commonly used to design and develop machinery will be used. These units consist of the U.S. Customary units inch, pound, and second, and the International System (SI) units meter, newton, and second. A complete description of units and their use

Table 3.1: Characteristics used in developing fluid power equations

Symbol	Characteristic	U.S. Customary	SI Units
ρ	Density	$lb_f \cdot s^2/in.^4$	$N \cdot s^2/m^4$ or kg/m^3
μ	Viscosity	$lb_f \cdot s/in.^2$	$N \cdot s/m^2$
Θ	Temperature	$^\circ F$	$^\circ C$
p	Pressure	$lb_f/in.^2$	N/m^2
X	Dimension	in.	m
Y	Dimension	in.	m
Z	Dimension	in.	m
t	Time	s	s

can be found in fluid mechanics texts and fluid power reference manuals [2, 3].

The characteristics generally used to model a fluid power system are shown in Table 3.1 and are derived from the fundamental units discussed in Chapter 2. The equations that are most useful for describing the operation of hydraulic power systems will now be introduced. The general equations from fluid mechanics that are useful in modeling an oil hydraulic machine follow:

$$\rho = \rho(p, \Theta) \tag{3.1}$$

$$\mu = \mu(p, \Theta) \tag{3.2}$$

Equations 3.1 and 3.2 express density and viscosity as a function of pressure and temperature. Density and viscosity, however, are generally expressed as constants at a particular temperature for most hydraulic power machinery analysis. The exact form for the relationships can be found in references on hydraulic oil [1, 4, 5].

The law of conservation of mass may be stated as:

$$\sum Q_{in} = \sum Q_{out} \tag{3.3}$$

or

$$\sum \dot{m}_{in} = \sum \dot{m}_{out} \tag{3.4}$$

It should be recognized that Equations 3.3 and 3.4 are not identical. Generally, the density of common hydraulic fluids does not change much with pressure so a conservation of volumetric flow approach is adequate for most hydraulic circuit analysis. Note that this last statement would not be true for pneumatic circuits.

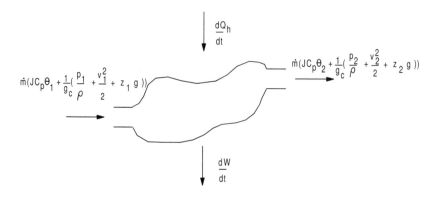

Figure 3.1: Energy conservation in a flow-through system.

3.3 CONSERVATION OF FLOW AND ENERGY

Fluid flow, Q, is expressed in units of volume per second or as $\dot{m}$ in units of mass per second. These equations are also referred to as the *continuity principle*. The equations state that all fluid or mass flowing into a defined system must be accounted for mathematically. The equations can be used in a variety of ways in hydraulic machinery analysis.

The law of conservation of energy or first law of thermodynamics is useful in describing the function of energy in a hydraulic power system. Several expressions, which may be used to describe elements or components within a system, may be derived from this law. The conservation of energy principle is shown schematically in Figure 3.1 and may be expressed formally as:

$$\frac{dQ_h}{dt} - \frac{dW}{dt} + \dot{m}\left(JC_P\Theta_1 + \frac{1}{g_c}\left(\frac{p_1}{\rho} + \frac{v_1^2}{2} + z_1 g\right)\right) =$$

$$\dot{m}\left(JC_P\Theta_2 + \frac{1}{g_c}\left(\frac{p_2}{\rho} + \frac{v_2^2}{2} + z_2 g\right)\right) \qquad (3.5)$$

The energy equation (Equation 3.5) is very useful for establishing design values for characteristics at any point in a fluid power system. Equation 3.5 may be used with any consistent units. The mechanical equivalent of heat, J, will have a value of 1 N · m/J for SI units, and 9339 lb$_f$ · in./Btu when

Table 3.2: Characteristics for the energy equation

Symbol	Characteristics	U.S. Customary Units	SI Units
dQ_h/dt	Heat flow	$\text{lb}_f \cdot \text{in.}/\text{s}$	$\text{N} \cdot \text{m}/\text{s}$
$P = dW_x/dt$	Power	$\text{lb}_f \cdot \text{in.}/\text{s}$	$\text{N} \cdot \text{m}/\text{s}$
$\dot{m}$	Mass flow	lb_m/s	kg/s
g_c	Mass conversion	$\text{lb}_m/(\text{lb}_f \cdot \text{s}^2/\text{in.})$	kg/kg
C_p	Specific heat	$\text{Btu}/\text{lb}_m \cdot {}^\circ\text{F}$	$\text{J}/\text{kg} \cdot {}^\circ\text{C}$
p	Pressure	$\text{lb}_f/\text{in.}^2$	$\text{N}/\text{m}^2 \ (\text{Pa})$
v	Fluid velocity	$\text{in.}/\text{s}$	m/s

U.S. Customary units with the inch as the length dimension are used. Although the potential energy term zg has been shown, this term can normally be ignored in fluid power system analysis.

The mass unit used in Equation 3.5 deserves some discussion. It will be observed that each term in the equation has the units of energy per unit time. If mechanical, this will represent power. Let us examine the group $(\dot{m}/g_c)p/\rho$. Working in the MLT dimensional system, this group has the dimensions ML^2T^{-3}. In the FLT dimensional system, this converts to FLT^{-1}, that is work per unit time as expected. Thus the mass flow rate unit chosen must be compatible. In a strict set of compatible units where mass, length, and time are slugs, feet, and seconds then the mass flow rate would be slug/s or the flow rate would be $(\text{lb}_f.\text{s}^2/\text{in.})/\text{s}$ where inches are the linear dimension. In SI units, the units would be kg/s. In U.S. Customary units mass flows are almost always expressed in lb_m/s. The introduction of the mass conversion constant g_c in Equation 3.5 allows the same equation to be used for U.S. Customary and SI units. Note that for U.S. Customary units, the mass flow rate, $\dot{m}$, will be lb_m/s for both *feet* and *inches* as the linear dimensions. In SI the mass flow rate will be kg/s. The mass conversion constant, g_c, will be 32.17 lb_m/slug for feet, 386.04 $\text{lb}_m/(\text{lb}_f \cdot \text{s}^2/\text{in.})$ for inches in U.S. Customary units and 1 kg/kg for SI units.

Tables 3.1 and 3.2 list the units for characteristics used in Equation 3.5. Average values for specific heat, C_p, for hydraulic oil, are 0.5 $\text{Btu}/\text{lb}_m \cdot {}^\circ\text{F}$ in U.S. Customary units and 2100 $\text{J}/\text{kg} \cdot {}^\circ\text{C}$ in SI units. If more precise values are needed for a particular fluid, manufacturer's data must be used.

Although actual fluid power machines may either add or extract power, Figure 3.1 is shown with the traditional thermodynamic convention where positive heat is counted as supplied to the system and positive work counts as the system doing work on the external environment. Heat is not used to drive fluid power machines. Power input to fluid power machines usually consists of energy from a rotating shaft. This power is generally added with an electric motor or an internal combustion engine driving a pump. In a fluid power system, heat always flows from the system. To maintain a relatively constant operating temperature, heat exchangers must often be installed in a system to provide sufficient heat removal. The energy equation may be used to determine the required capacity for an appropriate heat exchanger.

3.4 FRICTION LOSSES IN PIPES AND FITTINGS

Many equations that are useful for hydraulic power system analysis may be derived from Equation 3.5. Friction losses that occur during fluid flow, however, generally have to be evaluated from a mixture of analytical and empirical results. Fluid flow may by dominated by inertia or viscosity, depending on the value of Reynolds number, that is, the ratio of fluid inertia forces to fluid viscous forces. Reynolds number is usually expressed as:

$$Re = \frac{\rho v d}{\mu} = \frac{4\rho Q}{\pi \mu d} \tag{3.6}$$

for flow in a circular pipe.

When Reynolds number is below 2000 it is likely that laminar flow exists. In this regime, flow is smooth, steady, and occurs in parallel layers. When the Reynolds number exceeds 4000, however, the flow takes on an unsteady character with vortices crossing the fluid stream. This regime is termed turbulent. For Reynolds numbers varying between 2000 and 4000 we have transitional flow.

The theory for laminar flow has been highly developed. Starting with Equation 2.1 and assuming steady flow and constant velocity, then the pressure drop in hydraulic oil flowing along steel tubes or flexible hoses with circular cross section is:

$$\Delta p = \frac{f \ell \rho v^2}{2d} \tag{3.7}$$

where theoretical analysis shows that Equation 3.7 may be written as:

$$\Delta p = \frac{128 \mu \ell Q}{\pi d^4} \tag{3.8}$$

hence the fluid friction factor, f, is:

$$f = \frac{64}{Re} \tag{3.9}$$

For turbulent flow, a theoretical analysis of flow resistance has not yet been possible. It was established empirically by Blasius, however, that for smooth pipe interiors we may express the fluid friction factor approximately as:

$$f = \frac{0.3164}{Re^{0.25}} \tag{3.10}$$

Thus the pressure drop along a smooth pipe where turbulent flow is taking place may be calculated from:

$$\Delta p = \frac{f \ell \rho v^2}{2d} = 0.242 \frac{\ell \mu^{0.25} \rho^{0.75} Q^{1.75}}{d^{4.75}} \tag{3.11}$$

The friction factor is not well defined in the transition region, however it may be approximated as:

$$f = 0.0000039 Re + 0.0242 \tag{3.12}$$

The reader may note that *roughness* has been ignored in the previous treatment of turbulent flow. In practice, the steel pipes and hoses used for conveying oil may be treated as hydraulically smooth with little loss of accuracy.

Equations for pressure drop in *noncircular* passages can be found in many texts and handbooks [1, 6]. For *turbulent* flow in closed noncircular passages, the equations above can be used with an equivalent hydraulic diameter, d_h. The hydraulic diameter may be expressed as:

$$d_h = \frac{4A}{S} \tag{3.13}$$

In this equation A is the flow section area and S is the flow section perimeter. Equation 3.13 cannot be used for flow in noncircular passages in the laminar and transitional regions, other expressions from the cited references must be used. Flow is generally considered laminar if the Reynolds number, Re, is between 0 and 2000.

Pressure losses due to geometric factors in flow passages (e.g. fittings) may be determined with the equation:

$$\Delta p = K_h \frac{\rho v^2}{2} = K_h \frac{\rho Q^2}{2A^2} \tag{3.14}$$

The characteristic, K_h, is a geometry factor and is dependent on the flow passage shape. Values for the factor can be found in many references [1,

6]. Equations 3.7 through 3.14 are in general use in fluid power analysis, however, for critical applications independent laboratory studies may be necessary to obtained desired accuracy.

3.5 BASIC COMPONENT EQUATIONS

The main components for a fluid power system are pumps, motors, control valves, actuators, heat exchangers, accumulators, filters, and connecting lines. The operation of accumulators involves use of the fundamental gas laws. The influence of flow lines and passages may be determined with the use of Equations 3.7 through 3.14 as given above. Descriptive equations for other components can be derived from the energy equation.

The basic equation for pumps and motors is (for more information see Chapter 8):

$$P = \Delta p Q = T\omega \tag{3.15}$$

The quantity ΔpQ, is called hydraulic or fluid power. Since fluid pumps and motors are devices that utilize energy, Equation 3.15 can be derived from the energy equation. For an ideal situation, heat flow and temperature changes may be neglected. Also, if velocities v_1 and v_2 are assumed to be equal, Equation 3.15 will result. Power is expressed as horsepower in U.S. Customary units and watts in SI units. Flow is expressed as:

$$Q = D_m\omega \tag{3.16}$$

Displacement for a motor, D_m, or pump, D_p, is the volume of flow per unit shaft rotation. Basic units for displacement are in.^{2}/rad or m^3/rad. An expression for torque, T, is derived from combining Equations 3.15 and 3.16:

$$T = D_m\Delta p \tag{3.17}$$

Units for shaft speed, ω, are rad/s and units for torque are lb$_f$ · in. and N · m.

The above equations are written for ideal pumps and motors that are 100% efficient. In actual use the equations must be corrected with efficiency values. Efficiency terms include volumetric, η_V, mechanical, η_M, and, overall, η_O values. The relationship of efficiency values to other characteristics is shown in Table 3.3. These equations are discussed in more detail in Chapter 8.

Overall efficiency, η_O, is expressed as,

$$\eta_O = \eta_V\eta_M \tag{3.18}$$

A great variety of valves are used to control fluid power systems (Chapter 7). The elementary orifice equation may be used to describe flow

Table 3.3: Pump and motor efficiencies

Efficiency	Pump	Motor
η_O	$\Delta p Q / T\omega$	$T\omega / \Delta p Q$
η_V	$Q_{out} / D_p \omega$	$D_m \omega / Q_{in}$
η_M	$D_p \Delta p / T_{in}$	$T_{out} / D_m \Delta p$

through most valves. The equation can be developed from the energy equation [1, 3]. The basic equation is:

$$Q = C_d A \sqrt{\frac{2\Delta p}{\rho}} \tag{3.19}$$

The discharge coefficient, C_d, must be applied to the expression because of assumptions made in the development of the equation. The flow coefficient may be a variable, dependent on valve position. Experimental values are available in fluid power literature [1, 6, 7]. One of the difficulties encountered in predicting a value for the flow coefficient is that the flow through the orifice is unsteady and the walls are often moving. Experimental work is, however, available for flow in annular orifices under dynamic conditions with one wall moving and the other stationary [8]. The work shows that the difference in values of flow coefficient for dynamic and static conditions is generally less than 7%. An average value of C_d about 0.62 is often used.

For a spool valve, the flow area, A, is sometimes expressed as [1]:

$$A = w_x x \tag{3.20}$$

The term, w_x, is called the *flow area gradient* and it is expressed in units of in.^{2}/in. or m^2/m [1]. The flow gradient may be nonlinear for some valves. The dimension, x, represents the distance that the valve has opened for flow to occur.

In addition to motors, work is accomplished in fluid power machines with the use of actuators. These items are also referred to as rams or cylinders. They produce a force, F, that is basically stated as:

$$F = p_1 A_1 - p_2 A_2 \tag{3.21}$$

Linear actuators may have different areas operated on by the fluid on each side of the piston as shown by the use of A_1 and A_2 in Equation 3.21. Actuators may also be of a rotary type that produce torque, T, rather than a force. Rotary actuators are essentially fluid motors with limited shaft rotation.

Because efficiency of engineering systems is never 100%, energy loss is converted into heat that raises oil temperature. Thus heat exchangers are necessary for most fluid power systems. As described earlier, the energy equation may be used to determine the capacity of heat exchanger required.

Oil filters produce an undesirable pressure drop in a fluid system. They are necessary, however, to provide clean fluid. They are primarily selected in regard to the size of particles that must be removed from the fluid stream. Observe that filters are normally placed in suction or other low pressure lines.

Accumulators are used to store energy in a system. Energy is usually stored by compressing a coil spring or a bladder filled with nitrogen gas [9]. A volume of oil is then available at a particular pressure to do work as needed. The amount of energy available can be determined from a knowledge of the characteristics of the spring. If a gas type accumulator is used the gas laws from thermodynamics must be used. The equation:

$$p_1 V_1^n = p_2 V_2^n = p_3 V_3^n \tag{3.22}$$

describes the compressive process for a gas [7]. The characteristics p and V represent pressure and volume in the accumulator gas chamber. If the expansion and contraction of the gas takes place slowly very little heat is generated and recovery of the system to the original temperature is achieved. The process is then *isothermal* and the exponent n is equal to 1.0. If expansion and contraction take place rapidly, heat is generated and recovery to the initial temperature is not possible. The process then becomes *adiabatic* and the exponent n is equal to 1.4. Most gas type accumulators operate with a *polytropic* process, where the value of the exponent n falls between 1.0 and 1.4.

Use of Equations 3.1 to 3.22 form the basis for analysis of fluid power systems. The use of these equations along with use of the International Standards Organization, Fluid Power Systems and Components - Graphic Symbols, will be used in the chapters that follow to provide analysis of typical systems and components [2].

3.6 WORKED EXAMPLES

This section will present several worked examples showing the application of the equations presented earlier in this chapter. Most examples deal with transmissions, that is, pumps driving motors. There are also a valve example and an accumulator example.

In the transmission examples, the need for some form of heat exchange will be examined. Also, the power that is available at the motor shaft and the temperatures that develop in the fluid circuit are important to the design of the machine. The valve example will show that preliminary information on valve performance may often be obtained with little effort from steady state analyses before more complex time varying analyses are developed.

Some basic equations, peculiar to fluid power, are needed along with the energy equation to study the examples shown. These equations, some of which are shown in Chapter 2, are developed in several texts on hydraulics and will be repeated as needed [1, 4, 9].

3.6.1 Example: Oil Temperature Rise in a Hydrostatic Transmission System

Consider the system shown in Figure 3.2. For the purposes of this problem, most of the components shown in the diagram may be ignored. The individual components are treated in more detail in Chapter 10. Energy is transmitted from the pump to the motor, with some degradation to heat, and much of this heat is transferred out of the transmission by the flow of oil caused by the charge pump. The charge pump draws oil from a reservoir held at 100°F and at atmospheric pressure. The diameter of the line into the charge pump is the same as the line out.

Find the heat that must be removed by the heat exchanger and the temperature of the oil entering the heat exchanger. Work throughout in pound force, inch, second units.

In solving any problem, some assumptions must be made. In any system, some heat is lost from the motor and pump casings and from hoses. To simplify the example, we shall ignore these losses as minor and assume all heat rejected from the system is removed by the heat exchanger. Another implicit assumption is that oil is delivered to the reservoir from the heat exchanger at 100°F.

First establish the mass flow rate of oil through the charge pump using conversion factors derived from tables:

$$
\begin{aligned}
\dot{m} &= \rho_{OIL}Q \\
&= 62.4\frac{\text{lb}_\text{m}}{\text{ft}^3} \times 0.84 \times 2 \text{ USgal/min} \times 0.13368\frac{\text{ft}^3}{\text{USgal}} \times \frac{1}{60 \text{ s/min}} \\
&= 0.2335 \text{ lb}_\text{m}/\text{s}
\end{aligned}
$$

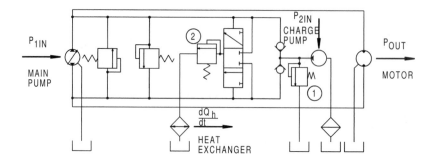

Figure 3.2: Heat discharge from a hydrostatic transmission.

Now calculate the oil density for U.S. Customary inch units:

$$\rho = \frac{1}{386.04 \text{ lb}_m/(\text{lb}_f \cdot \text{s}^2/\text{in.})}) \times 62.4\frac{\text{lb}_m}{\text{ft}^3} \times 0.84 \times \frac{1}{1728 \text{ in.}^3/\text{ft}^3}$$
$$= 78.58\text{E}-6 \text{ (lb}_f \cdot \text{s}^2/\text{in.})/\text{in.}^3$$

The statement of the problem indicated the performance of the charge pump, but did not give the power explicitly. An abbreviated form of Equation 3.5 can be used to calculate the power required to pump the given flow rate against the given pressure difference (hydraulic power). The statement of the problem indicates that this calculation will give 90% of the power input to the charge pump. There will be another 10% that is converted to heat because of friction and other pumping losses. The hydraulic power can be calculated from:

$$-\frac{dW}{dt} + \frac{\dot{m}\,p_1}{g_c\,\rho} = \frac{\dot{m}\,p_2}{g_c\,\rho}$$

express this equation as:

$$\frac{dW}{dt} = \frac{\dot{m}\,p_1}{g_c\,\rho} - \frac{\dot{m}\,p_2}{g_c\,\rho}$$

Table 3.4: Heat transfer example, U.S. Customary units

Characteristic	Value	Units
Power into the pump	20	hp
Power out of the motor	18	hp
Charge pump reservoir temperature	100	°F
Oil specific gravity	0.84	
Oil specific heat	0.5	$\mathrm{Btu/lb_m \cdot ° F}$
Pressure on low side of the transmission loop	200	$\mathrm{lb_f/in.}^2$
Charge pump flow	2	gal/min
Charge pump overall efficiency	90	%

Substituting in the values:

$$\frac{dW}{dt} = \frac{0.2335 \ \mathrm{lb_m/s}}{386.04 \ \mathrm{lb_m/(lb_f \cdot s^2/in.)}} \left(\frac{(0-200) \ \mathrm{lb_f/in.}^2}{78.58\mathrm{E}{-6} \ (\mathrm{lb_f \cdot s^2/in.)/in.}^3} \right)$$
$$= -1538 \ \mathrm{lb_f \cdot in./s}$$

Observe that the negative sign results from the thermodynamic convention used in Equation 3.5. Negative power means that power is being put into the system. This result matches the statement of the problem. The total rate of energy input, power plus heat, at 90% efficiency is given by:

$$\frac{dQ_h}{dt} = \frac{100\%}{90\%} \times 1538 \ \mathrm{lb_f \cdot in./s}$$
$$= 1709 \ \mathrm{lb_f \cdot in./s}$$

We can use dQ_h/dt because there is no work done by a pressure relief valve, so all the pressure energy ahead of a relief valve is converted to heat during the pressure drop through the valve.

We now have all the information necessary to calculate the heat that must be discharged by the heat exchanger. The statement of the problem indicated that 20 hp enters the system, but only 18 hp is delivered. Thus

2 hp is degenerated into heat because of losses in the pump and motor. Thus the heat rejected at the heat exchanger is:

$$\frac{dQ_h}{dt} = \frac{1}{9334 \text{ lb}_f \cdot \text{in./Btu}} \times (2 \text{ hp} \times 6600 \text{ lb}_f \cdot \text{in./hp} + 1709 \text{ lb}_f \cdot \text{in./s})$$

$$= \frac{1}{9334 \text{ lb}_f \cdot \text{in./Btu}} \times (14909 \text{ lb}_f \cdot \text{in./s}) = 1.597 \text{ Btu/s}$$

We may now calculate the temperature rise between the entry to the charge pump and the entry to the heat exchanger by again using an abbreviated form of Equation 3.5:

$$\frac{dQ_h}{dt} + \dot{m}C_p\Theta_1 = \dot{m}C_p\Theta_2$$

It should be noted that the mechanical equivalent of heat has been omitted because dQ_h/dt has just been evaluated in Btu/s. Hence:

$$\Theta_2 = \frac{1.597 \text{ Btu/s} + 0.2335 \text{ lb}_m/\text{s} \times 0.5 \text{ Btu/lb}_m \cdot^\circ \text{F} \times 100^\circ \text{F}}{0.2335 \text{ lb}_m/\text{s} \times 0.5 \text{ Btu/lb}_m \cdot^\circ \text{F}}$$

$$= 113.7^\circ \text{F}$$

At this point the reader should ask if providing a heat exchanger that could handle 1.597 Btu/s would be adequate for this system. The answer is, probably not. The designer should consider the expected operating conditions for the system. If there were a reasonable possibility that the motor could be stalled, then the heat exchanger should provide protection for a worst case situation. In such a situation, all 20 hp being delivered to the pump would appear as heat because a main relief valve would allow all the high pressure side oil to flow into the low pressure side. Under these conditions, the heat exchanger ought to be able to handle a little over ten times the heat transfer just calculated.

3.6.2   Example: A Pump Driving a Motor

Figure 3.3 denotes a simple fluid power system. In this system, a fluid pump provides flow to drive a fluid motor. Pressure develops in the flow line because of flow restrictions and the power that is taken from the motor at its shaft. Much information can be gained from the analytical study of this system under a variety of possible operating conditions. Basic information is needed in regard to steady-state operation of the system.

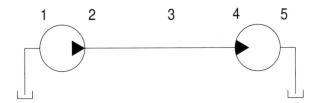

Figure 3.3: Fluid power pump used to drive a motor.

The system described in Figure 3.3 shows a fluid pump driving a motor. The components are connected with steel tubing. Useful design information on this type of system can be obtained with the energy equation. The equation can be applied between any two points on the circuit. Therefore, it is advantageous to identify key points on the figure with appropriate numbers.

The characteristics for the example in U.S. Customary units are given in Table 3.5. Observe that the fluid velocities are: $v_1 = v_2$ in./s, and $v_4 = v_5$ in./s.

1. Determine pump flow, Q, and pressures, p_2 and p_4, for the given conditions.

2. Determine motor output power, hp.

3. Determine the amount of heat flow in Btu/s that must be removed between points 2 and 4 to make temperature, Θ_4 equal to Θ_2.

Flow, Q, is determined with use of Equation 3.16:

$$
\begin{aligned}
Q & = D_p \omega = D_p n \frac{2\pi}{60} \\
& = 0.1 \frac{\text{in.}^3}{\text{rad}} \left(2000 \frac{\text{rev}}{\text{min}}\right) \left(2\pi \frac{\text{rad}}{\text{rev}}\right) \left(\frac{1}{60} \frac{\text{min}}{\text{s}}\right) \\
& = 20.9 \frac{\text{in.}^3}{\text{s}}
\end{aligned}
$$

Mass flow is expressed as:

$$
\dot{m} = Q\rho = 20.9 \frac{\text{in.}^3}{\text{s}} \left(0.03 \frac{\text{lb}_m}{\text{in.}^3}\right) = 0.63 \frac{\text{lb}_m}{\text{s}}
$$

Table 3.5: Pump power example, U.S. Customary units

Characteristic	Value	Units
Pump shaft power input, P	7	hp
Pump shaft speed, n	2000	rpm
Pump efficiency, η_{pO}	100	%
Pump displacement, D_p	0.1	in.^{3}/rad
Pressure at point 1, p_1	0	lb$_f$/in.^{2}
Pressure at point 5, p_5	100	lb$_f$/in.^{2}
Fluid viscosity, μ	3.9	lb$_f \cdot$ s/in.^{2}
Fluid density, ρ	0.03	lb$_m$/in.^{3}
Tube length from point 2 to 3, ℓ_2	30	in.
Tube length from point 3 to 4, ℓ_3	10	in.
Tube diameter from point 2 to 3, d_3	0.35	in.
Tube diameter from point 3 to 4, d_3	0.18	in.

To use Equation 3.5, ρ must be expressed in lb$_f \cdot$ s^2/in.^{4} as was discussed in Chapter 2:

$$\rho = \frac{0.03 \text{ lb}_m/\text{in.}^3}{386 \text{ lb}_m/(\text{lb}_f \cdot \text{s}^2/\text{in.})} = 77.7\text{E}{-}6 \text{ lb}_f \cdot \text{s}^2/\text{in.}^4$$

For 100% pump efficiency it may be assumed that temperatures, Θ_1, and, Θ_2, are equal and that heat flow is zero between points 1 and 2. Pressure, p_2, may then be determined by arranging Equation 3.5 as follows, and applying the known characteristic values:

$$
\begin{aligned}
p_2 &= \frac{g_c \rho}{\dot{m}} \frac{dW_x}{dt} + p_1 \\
&= \frac{(386 \text{ lb}_m/(\text{lb}_f \cdot \text{s}^2/\text{in.}^4))(77.7\text{E}{-}6 \text{ lb}_f \cdot \text{s}^2/\text{in.}^4)}{0.63 \text{ lb}_m/\text{s}} \\
&\quad \times (7 \text{ hp}) \left(6600 \frac{\text{lb}_f \cdot \text{in./s}}{\text{hp}} \right) + 0
\end{aligned}
$$

$$= 2200 \; \frac{\text{lb}_f}{\text{in.}^2}$$

Fluid velocities in the flow lines between point 2 and 3, and point 3 and 4 are:

$$v_2 = \frac{Q}{d_2^2 \pi/4} = \frac{20.9 \; \text{in.}^3/\text{s}}{0.35^2 \pi/4 \; \text{in.}^2} = 217.2 \; \frac{\text{in.}}{\text{s}}, \quad v_4 = \frac{20.9}{0.18^2 \pi/4} = 821.3 \; \frac{\text{in.}}{\text{s}}$$

Pressure drop between points 2 and 4 is a result of the fluid friction in the two sections of line and the sudden reduction in line size at point 3. The fluid friction factors, f, are dependent on the values of the Reynolds number. The pressure drop at point 3 occurs because of the sudden reduction in line size. A loss factor, K_h, must be applied at this point. The Reynolds numbers for line sections $2 \rightarrow 3$, and $3 \rightarrow 4$, as expressed by Equation 3.6 are:

$$Re = \frac{\rho v d}{\mu} = \frac{(77.7\text{E}{-}6 \; \text{lb}_f \cdot \text{s}^2/\text{in.}^4)(217.2 \; \text{in./s})(0.35 \; \text{in.})}{3.9\text{E}{-}6 \; \text{lb}_f \cdot \text{s}/\text{in.}^2} = 1515$$

$$Re = \frac{(77.7\text{E}{-}6)(831.3)(0.18)}{3.9\text{E}{-}6} = 2946$$

The Reynolds numbers indicate that flow is laminar in the first section of line, and in the transition region in the second section. The fluid friction factors may then be determined by use of Equations 3.9 and 3.12:

$$f = \frac{64}{Re} = \frac{64}{1515} = 0.042$$

$$f = 4.0\text{E}{-}6 Re + 0.024 = 4.0\text{E}{-}6(2946) + 0.024 = 0.036$$

Pressure drop at point 3 may be determined with a geometric loss factor and the use of Equation 3.14:

$$K_h = 0.5 \left(1 - \frac{d_3^2}{d_2^2} \right) = 0.5 \left(1 - \frac{0.18^2}{0.35^2} \right) = 0.37$$

Total pressure drop between points 2 and 4 may be expressed as:

$$\Delta p = f \frac{\ell_2}{d_2} \rho \frac{v_2^2}{2} + K_h \rho \frac{v_4^2}{2} + f \frac{\ell_3}{d_3} \rho \frac{v_4^2}{2}$$

$$\Delta p = 0.042 \left(\frac{30 \; \text{in.}}{0.35 \; \text{in.}} \right) (77.7\text{E}{-}6 \; \text{lb}_f \cdot \text{s}^2/\text{in.}^4) \left(\frac{(217.2 \; \text{in./s})^2}{2} \right)$$

$$+0.37(77.7\mathrm{E}{-}6\mathrm{lb_f} \cdot \mathrm{s^2/in.^4}) \left(\frac{(821.3\ \mathrm{in./s})^2}{2} \right)$$

$$+0.036 \left(\frac{10\ \mathrm{in.}}{0.18\ \mathrm{in.}} \right) (77.7\mathrm{E}{-}6\ \mathrm{lb_f} \cdot \mathrm{s^2/in.^4}) \left(\frac{(821.3\ \mathrm{in./s})^2}{2} \right)$$

$$= \quad 6.6 + 9.7 + 52.4 = 68.7\ \frac{\mathrm{lb_f}}{\mathrm{in.^2}}$$

Therefore, to complete part 1 of this example:

$$p_4 = p_2 - \Delta p = 2200 - 68.7 = 2131.3\ \frac{\mathrm{lb_f}}{\mathrm{in.^2}}$$

Motor output power for part 2 is:

$$
\begin{aligned}
P &= (p_4 - p_5)Q \\
&= (2131.3 - 100)\ \frac{\mathrm{lb_f}}{\mathrm{in.^2}} \left(20.9\ \frac{\mathrm{in.^3}}{\mathrm{s}} \right) \left(\frac{1}{6600}\ \frac{\mathrm{hp}}{\mathrm{lb_f} \cdot \mathrm{in./s}} \right) \\
&= 6.4\ \mathrm{hp}
\end{aligned}
$$

The loss in system power is due to the pressure losses in the lines and the return pressure in the motor outlet line. To maintain temperature Θ_4 equal to Θ_2 for part 3, the required heat flow between points 2 and 4 may be determined from the energy equation. Equation 3.5 may be arranged as follows:

$$\frac{dQ_h}{dt} = \frac{\dot{m}}{g_c} \left(\frac{p_2 - p_4}{\rho} + \frac{v_2^2 - v_4^2}{2} \right)$$

$$
\begin{aligned}
\frac{dQ_h}{dt} &= \frac{0.63\ \mathrm{lb_m/s}}{386\ \mathrm{lb_f} \cdot \mathrm{s/in.}} \\
&\quad \times \left(\frac{(2200 - 2131.3)\ \mathrm{lb_f/in.^2}}{77.7\mathrm{E}{-}6\ \mathrm{lb_f} \cdot \mathrm{s^2/in.^4}} \right. \\
&\quad \left. + \frac{(217.2^2 - 821.3^2)\ \mathrm{in.^2/s^2}}{2} \right) \\
&\quad \times \left(\frac{1}{9339}\ \frac{\mathrm{Btu}}{\mathrm{lb_f} \cdot \mathrm{in.}} \right) \\
&= 0.10\ \frac{\mathrm{Btu}}{\mathrm{s}}
\end{aligned}
$$

Table 3.6: Pump power example, SI units

Characteristics	Value	Units
Pump shaft power input, P	5.2	kW
Pump shaft speed, n	2000	rpm
Pump efficiency, η_{pO}	100	%
Pump displacement, D_p	1.64	cm^3/rad
Pressure at point 1, p_1	0.0	Pa
Pressure at point 5, p_5	6.9	MPa
Fluid viscosity, μ	27.0E-3	Pa·s
Fluid density, ρ	832.0	kg/m^3
Tube length from point 2 to 3, ℓ_2	762.0	mm
Tube length from point 3 to 4, ℓ_3	254.0	mm
Tube diameter from point 2 to 3, d_3	8.89	mm
Tube diameter from point 3 to 4, d_3	4.57	mm

3.6.3 Example: Using International System Units (SI)

The system described in Figure 3.3 can also be analyzed with use of the SI system of units. The converted characteristics from Table 3.5 are shown in Table 3.6. The same general equations from Chapter 2 may be used.

1. Determine pump flow, Q, and pressures, p_2 and p_4, for the given conditions.

2. Determine motor output power, kW.

3. Determine the amount of heat flow in J/s that must be removed between points 2 and 4 to make temperature Θ_4 equal to Θ_2.

The system can be analyzed with the same steps used for the US Customary units solution. Flow, Q, is determined with use of Equation 3.16:

$$Q = D_p\omega = D_p n \frac{2\pi}{60}$$

$$= 1.64 \frac{cm^3}{rad} \left(\frac{1}{100^3} \frac{m^3}{cm^3} \right) \left(2000 \frac{rev}{min} \right) \left(2\pi \frac{rad}{rev} \right) \left(\frac{1}{60} \frac{min}{s} \right)$$

$$= 0.343E{-}3 \frac{m^3}{s} = 0.343 \frac{L}{s}$$

Mass flow may be expressed as follows:

$$\dot{m} = Q\rho = 0.343E{-}3 \frac{m^3}{s} \left(832 \frac{kg}{m^3} \right) = 0.285 \frac{kg}{s}$$

For 100% pump efficiency it may be assumed that temperatures, Θ_1 and Θ_2, are equal and that heat flow is zero between points 1 and 2. Pressure, p_2, may then be determined by arranging Equation 3.5 as follows, and applying the known characteristic values:

$$
\begin{aligned}
p_2 &= \frac{g_c \rho}{\dot{m}} \frac{dW_x}{dt} + p_1 \\
&= \frac{(1\ kg/kg)(832\ kg/m^3)}{0.285\ kg/s} (5.2\ kW) \left(1000 \frac{N \cdot m/s}{kW} \right) + 0 \\
&= 15.2E{+}6 \frac{N}{m^2}
\end{aligned}
$$

Fluid velocities in the flow lines between point 2 and 3, and point 3 and 4 are:

$$v_2 = \frac{Q}{d_2^2 \pi/4} = \frac{0.343E{-}3\ m^3/s}{0.00889^2 (\pi/4)\ m^2} = 5.52\ m/s$$

$$v_4 = \frac{0.343E{-}3}{0.00457^2 \pi/4} = 20.91\ m/s$$

Pressure drop between points 2 and 4 is a result of the fluid friction in the two sections of line and the sudden reduction in line size at point 3. The fluid friction factors, f, are dependent on the values of the Reynolds number. The pressure drop at point 3 occurs because of the sudden reduction in line size. A loss factor, K_h, must be applied at this point. The Reynolds numbers for line sections $2 \rightarrow 3$, and $3 \rightarrow 4$, as expressed by Equation 3.6 are:

$$Re = \frac{\rho V d}{\mu} = \frac{(832\ N \cdot s^2/m^4)(5.52\ m/s)(0.00889\ m)}{27E{-}3\ N \cdot s/m^2} = 1512$$

$$Re = \frac{832 \times 20.91 \times 0.00457}{27E{-}3} = 2945$$

The Reynolds numbers indicate that flow is laminar in the first section of line, and in the transition region in the second section. The fluid friction factors may then be determined by use of Equations 3.9 and 3.12:

$$f = \frac{64}{Re} = \frac{64}{1512} = 0.042$$

$$f = 4.0\text{E}{-}6 \times Re + 0.024 = 4.0\text{E}{-}6 \times 2945 + 0.024 = 0.036$$

Pressure drop at point 3 may be determined with a geometric loss factor and the use of Equation 3.14:

$$K_h = 0.5\left(1 - \frac{d_3^2}{d_2^2}\right) = 0.5\left(1 - \frac{0.00457^2}{0.00889^2}\right) = 0.37$$

Total pressure drop between points 2 and 4 may be expressed as:

$$\Delta p = f\frac{\ell_2}{d_2}\rho\frac{v_2^2}{2} + K_h\rho\frac{v_4^2}{2} + f\frac{\ell_3}{d_3}\rho\frac{v_4^2}{2}$$

$$
\begin{aligned}
\Delta p &= 0.042\left(\frac{0.762\text{ m}}{0.00889\text{ m}}\right)\left(832\text{ N}\cdot\text{s}^2/\text{m}^4\right)\left(\frac{(5.52\text{ m/s})^2}{2}\right) \\
&\quad +0.37\left(832\text{ N}\cdot\text{s}^2/\text{m}^4\right)\left(\frac{(20.91\text{ m/s})^2}{2}\right) \\
&\quad +0.036\left(\frac{0.254\text{ m}}{0.00457\text{ m}}\right)\left(832\text{ N}\cdot\text{s}^2/\text{m}^4\right)\left(\frac{(20.91\text{ m/s})^2}{2}\right) \\
&= 45,632 + 67,298 + 363,933 = 476,863\ \frac{\text{N}}{\text{m}^2}
\end{aligned}
$$

Therefore, to complete part 1:

$$p_4 = p_2 - \Delta p = 15,180,351 - 476,863 = 14,703,488\ \frac{\text{N}}{\text{m}^2}$$

Motor output power for part 2 is:

$$
\begin{aligned}
P &= (p_4 - p_5)\,Q \\
&= (14,703,488 - 689,655)\ \frac{\text{N}}{\text{m}^2}\left(0.343\text{E}{-}3\ \frac{\text{m}^3}{\text{s}}\right) \\
&\quad \times\left(1\frac{\text{W}}{\text{N}\cdot\text{m/s}}\right)\left(\frac{1}{1000}\frac{\text{kW}}{\text{W}}\right) \\
&= 4.81\text{ kW}
\end{aligned}
$$

To maintain temperature Θ_4 equal to Θ_2 for part 3, the required heat flow between points 2 and 4 may be determined from the energy equation. Equation 3.5 may be arranged as follows:

$$\frac{dQ_h}{dt} = \frac{\dot{m}}{g_c}\left(\frac{p_2 - p_4}{\rho} + \frac{v_2^2 - v_4^2}{2}\right)$$

$$\begin{aligned}
\frac{dQ_h}{dt} &= 0.285\ \frac{\text{kg}}{\text{s}} \\
&\quad \times \left(\frac{(15,180,351 - 14,703,488)\ \text{N/m}^2}{832\ \text{N}\cdot\text{s}^2/\text{m}^4} + \frac{(5.52^2 - 20.91^2)\ \text{m}^2/\text{s}^2}{2}\right) \\
&\quad \times \left(1\ \frac{\text{J}}{\text{N}\cdot\text{m}}\right) \\
&= 105.4\ \frac{\text{J}}{\text{s}}
\end{aligned}$$

Units in the result are correct, because the force unit, which has been named the newton (N), is defined from Newton's 2nd law as:

$$1\ \text{N} = (1\ \text{kg})(1\ \text{m/s}^2)$$

3.6.4 Example: Incorporating Pump and Motor Efficiency Values

The examples in Sections 3.6.2 and 3.6.3 were solved with assumed efficiencies of 100% for the pump and motor. In this next example, the problem will be solved again with realistic efficiencies for the pump and motor. This example relates to Figure 3.3 with the same characteristic values that were used in that example. Pressure and flow at point 2, the outlet of the pump, can be determined with use of the definitions for efficiencies listed in Table 3.7. Therefore, since:

$$\eta_{pV} = \frac{Q_{out}}{D_p\omega}$$

$$Q = D_p n \frac{2\pi}{60} \eta_{pV} = 0.1\,(2000)\left(\frac{2\pi}{60}\right)(0.94) = 19.7\ \frac{\text{in.}^3}{\text{s}}$$

$$\dot{m} = 19.7\,(0.03) = 0.59\ \frac{\text{lb}_\text{m}}{\text{s}}$$

Table 3.7: Pump power example with efficiencies, U.S. Customary units

Characteristic	Value	Units
Pump shaft power input, P	7	hp
Pump shaft speed, n	2000	rpm
Pump displacement, D_p	0.1	in.^{3}/rad
Pump volumetric efficiency, η_{pV}	94	%
Pump mechanical efficiency, η_{pM}	95	%
Motor volumetric efficiency, η_{mV}	94	%
Motor mechanical efficiency, η_{mM}	95	%
Pressure at point 1, p_1	0	lb$_f$/in.^{2}
Pressure at point 5, p_5	100	lb$_f$/in.^{2}
Fluid viscosity, μ	3.9	lb$_f$.s/in.^{3}
Fluid density, ρ	0.03	lb$_m$/in.^{3}
Tube length from point 2 to 3, ℓ_2	30	in.
Tube length from point 3 to 4, ℓ_3	10	in.
Tube diameter from point 2 to 3, d_3	0.35	in.
Tube diameter from point 3 to 4, d_3	0.18	in.

Also:

$$\eta_{pM} = \frac{D_p \Delta p}{T_{in}}$$

Since:

$$P = T_{in}\omega$$

$$T_{in} = \frac{P}{\omega} = \frac{(7 \text{ hp})(6600 \text{ (lb}_f \cdot \text{in./s)/hp})}{(2000 \text{ rev/min}) \, (2\pi \text{ rad/rev})(1/60 \text{ min/s})} = 220.6 \text{ lb}_f \cdot \text{in.}$$

Therefore:

$$p_2 = \frac{T_{in}}{D_p}\eta_{pM} + p_1 = \frac{220.6 \text{ lb}_f \cdot \text{in.}}{0.1 \text{ in.}^3/\text{rad}}(0.95) = 2095.7 \frac{\text{lb}_f}{\text{in.}^2}$$

The application of efficiency values to the pump has caused both the pump output flow and the pressure to be reduced. Other characteristic values in the solution must then be determined with the reduced values. Fluid flow velocities, Reynolds numbers, and fluid friction factors must be adjusted as follows:

$$v_2 = \frac{Q}{d_2^2\pi/4} = \frac{19.7}{0.35^2\pi/4} = 204.7 \frac{\text{in.}}{\text{s}}, \qquad v_4 = \frac{19.7}{0.18^2\pi/4} = 774.1 \frac{\text{in.}}{\text{s}}$$

The Reynolds numbers and fluid friction factors for line sections $2 \rightarrow 3$, and $3 \rightarrow 4$, are:

$$Re = \frac{\rho v d}{\mu} = \frac{(77.7\text{E}{-}6)(204.7)(0.35)}{0.39\text{E}{-}5} = 1428$$

$$Re = \frac{(77.7\text{E}{-}6)(774.1)(0.18)}{0.39\text{E}{-}5} = 2777$$

$$f = \frac{64}{Re} = \frac{64}{1428} = 0.045$$

$$f = 4.0\text{E}{-}6Re + 0.024 = 4.0\text{E}{-}6\,(2777) + 0.024 = 0.035$$

The loss factor, K_h, at point 3 in Figure 3.3 does not change because it is dependent entirely on geometry. Total pressure drop between points 2 and 4 then becomes:

$$\Delta p = f\frac{\ell_2}{d_2}\rho\frac{v_2^2}{2} + K_h\rho\frac{v_4^2}{2} + f\frac{\ell_3}{d_3}\rho\frac{v_4^2}{2}$$

$$\begin{aligned}
\Delta p &= 0.045\left(\frac{30}{0.35}\right)\left(\frac{0.03}{386}\right)\left(\frac{204.7^2}{2}\right) + 0.37\left(\frac{0.03}{386}\right)\left(\frac{774.1^2}{2}\right) \\
&\quad +0.035\left(\frac{10}{0.18}\right)\left(\frac{0.03}{386}\right)\left(\frac{774.1^2}{2}\right) \\
&= 6.3 + 8.6 + 45.3 = 60.2 \frac{\text{lb}_f}{\text{in.}^2}
\end{aligned}$$

Therefore, to complete part 1:

$$p_4 = p_2 - \Delta p = 2095.7 - 60.2 = 2035.5 \frac{\text{lb}_f}{\text{in.}^2}$$

Motor output power for part 2 may then be determined with the use of Equations 3.15, and 3.18, and motor efficiency information found in Table 3.3, and other material located in Chapter 8. Then:

$$\eta_o = \eta_V \eta_M = \frac{T\omega}{\Delta p Q} = \frac{P_{out}}{P_{in}}$$

Therefore:

$$
\begin{aligned}
P_{out} &= \Delta p Q \eta_V \eta_M = (p_4 - p_5)\, Q \eta_V \eta_M \\
&= (2035.5 - 100)\ \frac{\text{lb}_\text{f}}{\text{in.}^2} \left(19.7\ \frac{\text{in.}^3}{\text{s}}\right)(0.93)\,(0.94)\left(\frac{1}{6600}\ \frac{\text{hp}}{\text{lb}_\text{f}\cdot\text{in.}/\text{s}}\right) \\
&= 5.0\ \text{hp}
\end{aligned}
$$

Equation 3.5 may then be used to determine the heat flow for part 3 of the solution:

$$\frac{dQ_h}{dt} = \frac{\dot{m}}{g_c}\left(\frac{p_2 - p_4}{\rho} + \frac{v_2^2 - v_4^2}{2g}\right)$$

$$
\begin{aligned}
\frac{dQ_h}{dt} &= \frac{0.59\ \text{lb}_\text{m}/\text{s}}{386\ \text{lb}_\text{m}/(\text{lb}_\text{f}\cdot\text{s}^2/\text{in.})} \\
&\quad \times \left(\frac{(2095.7 - 2035.5)\ \text{lb}_\text{f}/\text{in.}^2}{77.7\text{E}{-}6\ \text{lb}_\text{f}\cdot\text{s}^2/\text{in.}^4} + \frac{(204.7^2 - 774.1^2)\ \text{in.}^2/\text{s}^2}{2}\right) \\
&\quad \times \left(\frac{1}{9339}\ \frac{\text{Btu}}{\text{lb}_\text{f}\cdot\text{in.}}\right) \\
&= 0.08\ \frac{\text{Btu}}{\text{s}}
\end{aligned}
$$

This equation describes the energy that is lost from the system, between points 2 and 4, because of fluid friction and geometry related losses. Energy is also lost at the pump and motor because these components are less than 100% efficient. This energy is lost to the machine's surroundings as heat flow. Power output for the pump in the examples in Sections 3.6.2 and 3.6.3 may be expressed as:

$$P = Q\Delta p = 20.9\ \frac{\text{in.}^3}{\text{s}}(2200 - 0)\ \frac{\text{lb}_\text{f}}{\text{in.}^2} = 45980\ \frac{\text{lb}_\text{f}\cdot\text{in.}}{\text{s}}$$

$$P = Q\Delta p = 19.7\ \frac{\text{in.}^3}{\text{s}}(2095.7 - 0)\ \frac{\text{lb}_\text{f}}{\text{in.}^2} = 41285\ \frac{\text{lb}_\text{f}\cdot\text{in.}}{\text{s}}$$

Power lost as heat flow is:

$$\frac{dQ_h}{dt} = \left((45980 - 41285)\ \frac{\text{lb}_\text{f}\cdot\text{in.}}{\text{s}}\right)\left(\frac{1}{9339}\ \frac{\text{Btu}}{\text{lb}_\text{f}\cdot\text{in.}}\right) = 0.50\ \frac{\text{Btu}}{\text{s}}$$

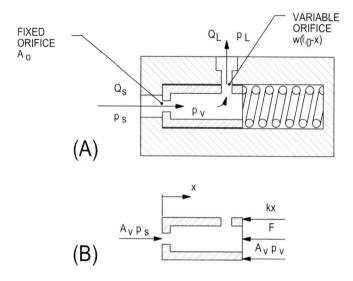

Figure 3.4: Physical configuration of a flow regulator valve.

3.6.5   Example: Performance of a Flow Regulator Valve

This next example will present use of the orifice flow equation, Equation 3.19.
Steady state analysis will not generally be adequate for determining the
complete performance of a valve because valves contain moving parts. An
initial steady state analysis, however, may be used to obtain an estimate of
valve characteristics.

A flow regulator valve is discussed in more detail in Chapters 7 and 13.
The valve serves to stabilize the flow to some device such as an actuator
as the system pressure upstream of the actuator changes. The physical
configuration of the valve is shown in Figure 3.4. There are two orifices in
series and the second will vary in size as the pressure drop across the spool,
$p_s - p_v$, changes. The spool has a preload force, F, and a spring force,
kx, that resist the pressure difference across the spool. The component
characteristics may be chosen so the flow through the valve remains nearly
constant over the expected change in upstream pressure, p_s.

Consider the valve with the characteristics shown in Table 3.8. Note
that for simplicity, the load pressure, p_L, has been shown as constant.

There are two characteristics that can be obtained from a steady state
analysis. The valve spool displacement, x, and the load flow through the

Table 3.8: Flow regulator valve characteristics

Characteristic	Value	Units
Spool cross section area, A_v	0.129E−3	m^2
Fixed orifice area, A_o	6.45E−6	m^2
Variable orifice area gradient, w	2.5E−3	m^2/m
Spring rate, k	2700	N/m
Orifice discharge coefficient, C_d	0.6	
Initial variable orifice opening, ℓ_0	4.0E−3	m
Fluid density, ρ	855	kg/m^3
Load pressure, p_L	12.0E+6	Pa

valve, Q_L. The designer of a circuit will need to know how these values vary as the upstream pressure on the valve, p_s, varies over some specified range. After determining these quantities, we shall use the energy equation, Equation 3.5, to determine the loss of pressure energy incurred when using the valve.

First determine the flow through the valve. The flow through the fixed orifice is given by:

$$Q_L = C_d A_o \sqrt{\frac{2(p_s - p_v)}{\rho}} \qquad (3.23)$$

and the flow through the variable orifice to the load by:

$$Q_L = C_d w(\ell_0 - x)\sqrt{\frac{2(p_v - p_L)}{\rho}} \qquad (3.24)$$

These two equations may be combined to yield an expression for p_v:

$$p_v = \frac{A_o^2 p_s + (w(\ell_0 - x))^2 p_l}{A_o^2 + (w(\ell_0 - x))^2}$$

Now examine the static equilibrium of the spool:

$$p_s A_v = p_v A_v + F + kx$$

Rearrange this equation to yield p_v:

$$p_v = \frac{p_s A_v - F - kx}{A_v}$$

The quantity p_v can be eliminated leaving one equation in the spool displacement x:

$$\frac{p_s A_v - F - kx}{A_v} - \frac{A_o^2 p_s - (w(\ell_0 - x))^2 p_L}{A_o^2 + (w(\ell_0 - x))^2} = 0 \qquad (3.25)$$

In this instance, Equation 3.25 could be solved explicitly by rearranging it in the form of a cubic polynomial in x. As a general rule, such effort is not warranted. Instead, the equation in its current form may be solved numerically. In this instance Visual Basic for Applications® within a Microsoft Excel® spreadsheet was used so that results could easily be presented as graphs. After determining the spool displacement, x, at a series of p_s values, the equation derived by combining Equations 3.23 and 3.24:

$$Q_L = C_d \frac{1}{\sqrt{1/A_o^2 + 1/(w(\ell_0 - x))^2}} \sqrt{\frac{2(p_s - p_L)}{\rho}} \qquad (3.26)$$

was used to generate companion values of Q_L. The results are shown in Figures 3.5 and 3.6. As claimed in the introduction to this example, the flow to the load, Q_L, is quite constant after the system pressure has risen sufficiently to achieve the needed value of p_v to achieve p_L. An attractive feature of this steady state analysis is that considerable information about the valve characteristics can be obtained with little effort by the designer. Values of preloads, spring rates, and the valve dimensions can quickly be changed.

The pressure energy loss through this valve can be calculated using a simplified version of the energy equation, Equation 3.5. The simplified form may be used by noting that the valve does no external work, $dW/dt = 0$, heat transfer to or from the valve body is small enough to be ignored, $dQ_h/dt \approx 0$, the input and output pipe diameters will be taken as equal, and the change in head, Δz, may be ignored. Thus:

$$JC_p\Theta_1 + \frac{1}{g_c}\left(\frac{p_s}{\rho}\right) = JC_p\Theta_2 + \frac{1}{g_c}\left(\frac{p_L}{\rho}\right) \qquad (3.27)$$

There are several ways in which the energy loss could be expressed. One logical approach would be to express it as a percentage of the pressure energy in the fluid entering the valve. The expression for this derived from Equation 3.27 is:

$$\text{Percentage loss} = 100 \times \frac{\dot{m}(1/g_c)((p_s - p_L)/\rho)}{Q_L p_s}$$

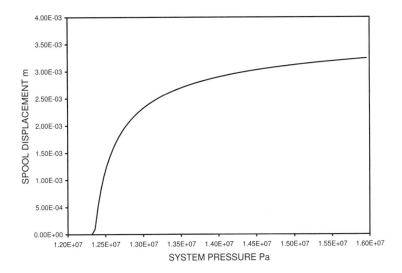

Figure 3.5: Flow regulator valve, spool displacement vs. upstream pressure, p_s.

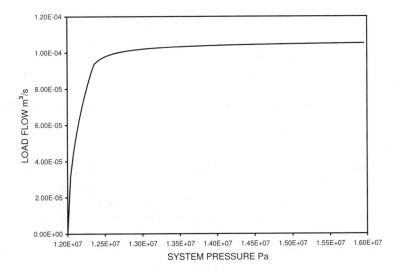

Figure 3.6: Flow regulator valve, load flow vs. upstream pressure, p_s.

Noting that $Q_L = \dot{m}/\rho$ and $g_c = 1\text{kg/kg}$ for SI units and evaluating the loss at only one pressure, $p_s = 15.0\text{E+}6$ Pa, then:

$$\begin{aligned}
\text{Percentage loss} &= 100 \times \frac{p_s - p_L}{p_s} \\
&= 100 \times \frac{15.0\text{E+}6 - 12.0\text{E+}6}{15.0\text{E+}6} \\
&= 20\%
\end{aligned}$$

In this instance, using the energy equation was overkill. The pressure energy loss through the valve is obviously Δp. Equation 3.27, however, can also be used to calculate the temperature rise of the fluid passing through the valve.

This example does show one prevalent disadvantage of fluid power. Each time fluid passes through a valve and a pressure drop results, there will be loss in power transmission efficiency. Often such inefficiency is tolerated because the fluid power system has other advantages such as compactness and the ability to place actuators where they are needed.

3.6.6   Example: Using an Accumulator

Some fluid power circuits operate on a uniform cyclic basis. An example might be an actuator driving a molding press. The duty cycle might have the sequence: fill the mold cavity, heat the polymer pellets, compress the liquefied polymer to form the item, cool, retract, and eject. The details do not concern us except that there will be a relatively long period during which the pump supplying the fluid to the actuator does no work. There will then be a short period in which the pump must deliver a relatively large flow at a high pressure. Such a system will require a large pump in order to satisfy the molding requirement, but this pump may be idle for a large portion of the cycle. This is the ideal system for installing an accumulator.

An accumulator is a heavy wall steel cylinder with domed ends that is divided into two chambers. The hydraulic fluid fills one chamber and nitrogen fills the other. The chambers are separated by a flexible polymer membrane.

When an accumulator is installed, a smaller pump can be used because this pump now charges the accumulator continuously. The large flow required to operate the press during the molding operation is provided by the compressed nitrogen forcing the fluid out of the accumulator.

Two systems, one without an accumulator and one with are shown in Figure 3.7. First consider Panel A, where there is no accumulator. During

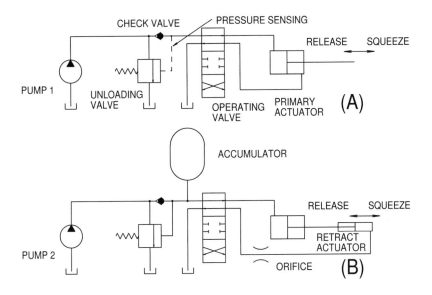

Figure 3.7: Actuator systems without and with an accumulator.

the nonmolding portion of the cycle, the pump should not experience any outlet pressure. This is achieved by using an unloading valve. The operating valve will be in the centered position where flow is blocked both on the pump side and the actuator side. Thus the pressure between the pump and the valve will rise rapidly. At a preset pressure, the unloading valve will open allowing the pump flow to reach the reservoir at atmospheric pressure. The check valve between the unloading valve and the operating valve will close, ensuring the pressure in this section remains high. and holds the unloading valve open. Under these conditions, the pump will absorb little power because the unloading valve is open and the pump operates against a neglible head. When the operating valve is opened to begin the molding stroke, the pressure in the line will fall below the unloading valve setting and the pump will provide full flow and pressure to the actuator.

There are two major reasons for using accumulators:

1. The cost of an accumulator is generally less than that of a large pump with its electric motor or engine.

Table 3.9: Molding press characteristics

Characteristic	Value	Units
Cycle time, t_{cyc}	120	s
Load force, F	0.1E+6	$\mathrm{lb_f}$
Retraction force	0	$\mathrm{lb_f}$
Power stroke, x_{max}	4.5	in.
Actuator velocity, v	3.0	in./s
Actuator pressure, p	3000	$\mathrm{lb_f/in.}^2$
Polytropic index, n	1.2	
Orifice coefficient, C_d	0.62	
Fluid density, ρ	75.0E−6	$\mathrm{lb_f \cdot s^2/in.}^4$

2. The peak power demand is significantly reduced.

The circuit shown in Figure 3.7 Panel B has three extra components. Obviously there is the accumulator, but there is also an orifice installed in the actuator discharge line and a retraction actuator. The orifice is necessary because the accumulator contents will still be at high pressure when the operating valve is moved to the retraction position. The orifice must be chosen to limit the retraction velocity. The retract actuator may be necessary to conserve the overall energy used in the cycle.

The characteristics of a molding press are given in Table 3.9. First consider the situation in Panel A. Given that an ideal, lossless situation is being analyzed, the pump power required is given by:

$$P = \frac{Fv}{6600} = \frac{0.1\mathrm{E}{+}6 \ \mathrm{lb_f} \times 3.0 \ \mathrm{in./s}}{6600 \ \mathrm{hp \cdot s/lb_f \cdot in.}} = 45.5 \ \mathrm{hp}$$

The actuator area is:

$$A = \frac{F}{p} = \frac{0.1\mathrm{E}{+}6 \ \mathrm{lb_f}}{3000 \ \mathrm{lb_f/in.}^2} = 33.3 \ \mathrm{in.}^2$$

This is equivalent to a diameter of 6.5 in. Now calculate the oil volume needed for the power stroke:

$$V_{sqz} = Ax = 33.3 \ \mathrm{in.}^2 \times 4.5 \ \mathrm{in.} = 150 \ \mathrm{in.}^3$$

The pump flow required will be:

$$Q = vA = 3.0 \ \mathrm{in./s} \times 33.3 \ \mathrm{in.}^2 = 100 \ \mathrm{in.}^3/s$$

Duration of the power stroke:

$$t_{sqz} = \frac{x_{max}}{v} = \frac{4.5 \text{ in.}}{3.0 \text{ in./s}} = 1.5 \text{ s}$$

The system characteristics stated that the retraction force was zero, so the work done on the press during the cycle is just the work done during the power stroke:

$$W = Fx = 0.1\text{E+6 lb}_f \times 4.5 \text{ in.} = 0.45\text{E+6 lb}_f \cdot \text{in.}$$

The accumulator and companion parts are now added to the system. Some additional system characteristics are also required. Let us assume that the minimum pressure required in the power stroke is 3000 $\text{lb}_f/\text{in.}^2$, consequently the accumulator must initially be charged to a pressure greater than this. Select the maximum pressure to be $p_1 = 3300$ $\text{lb}_f/\text{in.}^2$ and the pressure at the end of the power stroke will be $p_2 = 3000$ $\text{lb}_f/\text{in.}^2$ Let the gas volume at p_1 be V_1 and the volume at p_2 be V_2, noting $V_2 = V_1 + V_{sqz}$ where $V_{sqz} = 150$ in.^3 A polytropic process is being assumed so the relationship among these quantities is:

$$p_1 V_1^n = p_2 V_2^n$$

This relationship can be rearranged to yield V_1:

$$\begin{aligned} V_1 &= \left(\frac{p_2}{p_1}\right)^{1/n} V_2 \\ &= \left(\frac{p_2}{p_1}\right)^{1/n} (V_1 + V_{sqz}) \\ &= \frac{(p_2/p_1)^{1/n}}{1 - (p_2/p_1)^{1/n}} V_{sqz} \end{aligned}$$

Substituting the values just presented will yield V_1:

$$V_1 = \frac{(3000/3300)^{1/1.2}}{1 - (3000/3300)^{1/1.2}} 150 = 1815 \text{ in.}^3$$

Thus:

$$V_2 = V_1 + V_{sqz} = 1815 + 150 = 1965 \text{ in.}^3$$

The designer must now make some decisions. It is desirable that the actuator should retract rapidly. On the other hand, too rapid retraction with an accumulator pressure of 3000 $\text{lb}_f/\text{in.}^2$ available could cause damage.

The power stroke takes 1.5 s so it seemed reasonable to retract in 1 s. The next consideration is the oil volume expelled from the accumulator during retraction. This should be kept to a minimum because it represents unnecessary pump work. It is this requirement that motivated adding a separate retraction actuator in Figure 3.7. Whether this is necessary will now be discussed.

For Trial 1, we shall first consider retraction using the main actuator. Actuators come in two patterns. Rods may be attached to both ends of the piston so the areas operated on by the pressure are the same for both directions of piston motion. In the other pattern, there is only one rod so this side has less operating area than the other or cap side. The single rod design is the better choice for this application because the smaller area on the rod side means that, for a given displacement, the fluid volume on the rod side is less than that on the cap side. In this example, it has been assumed that the rod is half the diameter of the piston. The cap and rod areas are given by:

$$\text{Cap side area} \quad = \quad \frac{\pi d^2}{4}$$

$$\text{Rod side area} \quad = \quad \frac{\pi d^2}{4} - \frac{\pi d^2}{16}$$

$$= \quad \frac{3}{4}\frac{\pi d^2}{4}$$

Because the power and retraction strokes are the same length, the fluid expelled from the accumulator during the retraction stroke is:

$$V_{frtc} = 0.75 \times V_{sqz} = 0.75 \times 150 = 112.5 \text{ in.}^3$$

Hence the gas volume in the accumulator at the end of the retraction stroke will be:

$$V_3 = V_2 + V_{frtc} = 1964.6 + 112.5 = 2077 \text{ in.}^3$$

Applying the polytropic expression again will yield the pressure in the accumulator at the end of the retraction stroke:

$$p_3 = \left(\frac{V_2}{V_3}\right)^n p_2 = \left(\frac{1965}{2077}\right)^{1.2} 3000 = 2806 \text{ lb}_f/\text{in.}^2$$

The final part of the design is the selection of the orifice area so the retraction stroke does take 1 s. A generalized and rigorous approach would require development of a differential equation between pressure and time for varying flow through an orifice. In this specific example a more primitive approach gives quite satisfactory results. Although graphs of flow from or pressure in the accumulator during the retraction stroke are not linear, they are not strongly curved. If they are treated as linear, then the orifice area can be calculated based on a mean flow. Because the relation between flow and pressure through an orifice follows a square root law, first find:

$$p_{mean} = 0.5(\sqrt{p_2} + \sqrt{p_3}) = 0.5(\sqrt{3000} + \sqrt{2806}) = 2902 \text{ lb}_f/\text{in.}^2$$

The desired mean flow from the orifice during retraction can be obtained from:

$$Q_{mean} = \frac{V_{frtc}}{t_{rtc}} = \frac{112.5 \text{ in.}^3}{1 \text{ s}} = 112.5 \text{ in.}^3/\text{s}$$

Thus the desired orifice area can be found from rearrangement of Equation 3.19:

$$
\begin{aligned}
A_o &= \frac{Q_{mean}}{C_d} \sqrt{\frac{\rho}{2p_{mean}}} \\
&= \frac{112.5 \text{ in.}^3/\text{s}}{0.62} \sqrt{\frac{75.0\text{E}{-}6 \text{ lb}_f \cdot \text{s}^2/\text{in.}^4}{2 \times 2902 \text{ lb}_f/\text{in.}^2}} \\
&= 0.0206 \text{ in.}^2
\end{aligned}
$$

The pump power required to operate the press with an accumulator may be found from the total oil pumped into the accumulator, the time required to do this, and the maximum pressure that is experienced by the pump. Noting that the change in gas volume and the change in oil volumes are the same:

$$P = \frac{(V_3 - V_1)p_1}{(t_{cyc} - t_{sqz} - t_{trc})}$$

The cycle time for the press was given as 120 s in Table 3.9 and the power and retract strokes occupied a total of 2.5 s. Thus the pump must charge the accumulator from 2806 lb$_f$/in.^{2} to 3300 lb$_f$/in.^{2} in 117.5 s. The work require during a polytropic process is:

$$
\begin{aligned}
W &= \frac{1}{1 - n}(p_2 V_2 - p_1 V_1) \\
&= \frac{1}{1 - 1.2}(3300 \text{ lb}_f/\text{in.}^2 \times 1815 \text{ in.}^3 - \\
&\qquad 2806 \text{ lb}_f/\text{in.}^2 \times 2077 \text{ in.}^3) \\
&= -0.798\text{E}{+}6 \text{ lb}_f\cdot\text{in.}
\end{aligned}
$$

Table 3.10: Press example with accumulator, trial 1 calculated values

Characteristic	Value	Units
Retraction time, t_{rtc}	1.0	s
Final acc. pressure, p_3	2806	$lb_f/in.^2$
Init. acc. gas vol., V_1	1815	$in.^3$
End of power strk. acc. gas. vol., V_2	1965	$in.^3$
End of retract strk. acc. gas vol., V_3	2077	$in.^3$
Retract fluid vol., V_{rtc}	112.5	$in.^3$
Orifice area, A_o	20.6E$-$6	$in.^2$
Pump power, P	1.12	hp
Pump work, W	0.798E+6	$lb_f \cdot in.$

Table 3.11: Press example with accumulator, trial 2 calculated values

Characteristic	Value	Units
Retraction time, t_{rtc}	1.0	s
Final acc. pressure, p_3	2996	$lb_f/in.^2$
Init. acc. gas vol., V_1	1815	$in.^3$
End of power strk. acc. gas. vol., V_2	1965	$in.^3$
End of retract strk. acc. gas vol., V_3	1967	$in.^3$
Retract fluid vol., V_{rtc}	1.99	$in.^3$
Orifice area, A_o	0.359E$-$6	$in.^2$
Pump power, P	0.647	hp
Pump work, W	0.478E+6	$lb_f \cdot in.$

Note that the work calculated by the polytropic expression is negative because work is done on the gas by the oil and thermodynamically, this is negative. Compare this value with the work obtained for the direct pump method. The work require per cycle is nearly twice as great. Obviously the beneficial reduction in pump size has been partly offset by the increase in work requirement. The reason that the work requirement has increased so much is that the actuator was retracted by directing the accumulator into the rod side of the main actuator. This required a discharge of 122.5 in.^{3} of fluid from the accumulator and a reduction in accumulator pressure to 2806 lb_f/in.^{2}

The statement of the problem contained in Table 3.9, indicated that the retraction force was zero. Consequently the designer could select a much smaller retraction actuator that will reduce the amount of oil discharged by the accumulator during retraction. A benefit of this will be that the lowest pressure in the accumulator will be raised. For Trial 2, a retraction accumulator with a 0.75 in. bore is chosen, then the values in Table 3.11 are obtained. The calculation procedure follows that just presented for Trial 1 and will not be repeated here.

In closing, note that the final results obtained by reducing the retraction actuator are quite significant. The total oil volume that must be displaced by the pump has been significantly reduced, consequently both the pump size has been reduced (from 1.12 to 0.65 hp) and the work done during the cycle has also been reduced (0.798 $lb_f \cdot$ in. to 0.478 $lb_f \cdot$ in.). It will be observed that the orifice size is much smaller in Trial 2 because the needed retraction oil flow has been much decreased. The example was simplified by assuming that the force during retraction was zero. In practice this is unlikely to be the situation, although it is likely that the retraction force will be much less than the power stroke force. If there is a retraction force, the size of the retraction actuator must accommodate this force. Note that in the preceding development, the pressure downstream of the orifice was zero. If the retraction force is not zero, then this pressure will not be zero and the orifice must be sized for the appropriate pressure differential.

3.7 DISCUSSION

In the six examples presented in this chapter, the pump and motor systems were obviously steady state. The flow regulator valve and the accumulator were not steady state. A designer must exercise care when treating a moving system as if dynamics can be ignored. In the flow regulator example, the steady state analysis would be quite useful when the physical size of the valve components was being decided. The analysis could also give information on the range of load pressure that could be used for which the flow to the actuator would not change significantly. On the other hand, the steady state analysis does not give any information on possible oscillations of the valve spool. If these oscillation were sufficiently large, then the valve would close prematurely and large pressure excursions might occur in the supply line.

Dynamic analyses of this nature require introduction of $F = ma$ equations. A companion effect is the effect of system compliance, $dp/dt = (\beta/V)dV/dt$. When these effects are introduced, differential equations result and different solution techniques are required (Chapters 4 and 13). Similar comments can be made about the accumulator example. The analysis performed was perfectly satisfactory for the specified purpose, finding the power and energy required to drive the system. If the time vs. displacement response of the actuator during the power stroke were needed, more sophisticated dynamic analyses would be required.

There are two basic reasons for writing a mathematical model for a fluid power system. The primary benefit derived for establishing a model is to promote understanding in regard to system function. The model then allows evaluation of system operation, which provides the important second benefit. Computer software, both general purpose and that available specifically for fluid power analysis, allows easy changes in system variables. Therefore, establishment of a mathematical model allows study of the system for the complete range of expected characteristic variation.

Information on modeling is available from many sources. Most of the texts and handbooks specifically written on fluid power systems will provide useful information. Computer software specifically written for fluid power analysis usually contains helpful information. Some typical examples are listed in the references at the end of this chapter [1, 2, 4, 9, 10].

A mathematical model consists of one or more equations that define the interaction among selected system variables. The model may be structured to accomplish any desired analytical task. In the general case, some assumptions and simplifications will be necessary, which may limit the range of usefulness of the model. On the other hand, models are flexible and allow the user to incorporate as much complexity as desired. It then becomes possible to select those system characteristics or variables that have the greatest influence on system operation. A good model will ultimately lead to a better understanding of the system being studied. Also, it will define the magnitude of various characteristics to provide the desired system performance. The flexibility and overall usefulness of fluid power is very advantageous. Where possible, such aspects should be included in a model.

Available data and knowledge on a particular system will define the nature of the mathematical model. Results gained from mathematical analysis must always be compared to actual operation of hydraulic power systems. Comparison will increase confidence in mathematical modeling procedures. The establishment of a model will always lead to better understanding of system performance. Therefore, the time spent to write the model will be well rewarded.

PROBLEMS

3.1 A truck with a hydraulic boom is equipped with a hydraulic chain saw at the operator's basket as shown in the figure. The pump is mounted on the truck. The characteristics of the system are shown in the table. The temperature is constant in the system. All flow velocities in the system are equal.

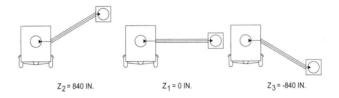

$Z_2 = 840$ IN. $Z_1 = 0$ IN. $Z_3 = -840$ IN.

Determine the power available at the motor for each of the three elevations Z_1, Z_2, and Z_3 given on the figure.

Characteristics of a system using a movable chain saw

Characteristic	Size	Units
Pump displacement, D_p	0.45	in.^{3}/rev
Pump outlet pressure, p_s	2100	lb$_f$/in.^{2}
Pump shaft speed	1200	rpm
Pressure loss between pump and motor, Δp	450	lb$_f$/in.^{2}
Oil density, ρ	0.03	lb$_m$/in.^{3}
Motor return pressure, p_d	0.0	lb$_f$/in.^{2}

3.2 A fluid power pump is used to drive a motor as shown in the figure. The components are connected with commercial steel tubing. De-

termine the correct pump speed, n, to provide a flow velocity, v, in
the line of 7.0 m/s. Determine the allowable torque output from the
motor to provide a factor of safety of 4 for the tube wall.

Characteristics of a pump and motor transmission system

Characteristic	Size	Units
Tube length, ℓ_L	1.2	m
Tube external diameter, d_o	12.0	mm
Tube wall thickness, ℓ_t	0.1	mm
Tube tensile strength, S_t	395	MPa
Oil viscosity, μ	11.5	mPa · s
Oil density, ρ	837	kg/m^3
Pump displacement, D_p	20.0	mL/rev
Pump inlet pressure, p_0	0.0	kPa
Pump volumetric efficiency, η_{vp}	95.0	%
Valve loss factor, K	10.0	
Motor displacement, D_m	37.0	mL/rev
Motor mechanical efficiency, η_{mm}	94.0	%
Motor outlet pressure, p_3	500	kPa

3.3 A hydraulic motor is used to drive the rear wheels of a truck through
a drive shaft. The truck is equipped with an accumulator that stores
the kinetic energy of the truck during deceleration.

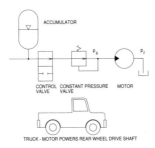

Determine the kinetic energy, KE, that can be stored in the accumulator if the truck is travelling at a velocity 70 kph and the energy is transferred to the accumulator with a pump as the truck decelerates to 0 kph. Determine the power, P, that is available to drive the truck if the energy from the previous part is used in a 30 s time interval. Determine the acceleration of the rear wheels caused by the pressure, p_s, applied to the motor. Determine the acceleration of the rear wheels caused by the pressure, p_s, applied to the motor when a load torque, T_L, of 175 N · m develops at the rear wheels. NOTE: T_L is the combined load torque for both rear wheels. Determine the acceleration of the rear wheels for the conditions given in the previous part, but with the mass of the truck, m, included in the calculation.

Simplified regenerative energy storage for a truck

Characteristic	Size	Units
System pressure, p_s	32.0	MPa
Motor return pressure, p_r	250.0	kPa
Motor displacement, D_m	55.0	mL/rev
Rear wheel diameter, d	900	mm
Truck mass, m	1700	kg
Moment of inertia, I_c	20.0	kg · m^2
(I_c is combined moment for		
2 rear wheels and drive shaft)		

3.4 A hydraulic lift consists of a large ram and a hand operated pump. Determine the pressure, p, and the force, F, that is needed to raise the mass, m. Determine the power, P, that is developed to raise the mass, m, a distance, h, during the given interval. Determine the number of strokes required to raise the mass, m, during the interval.

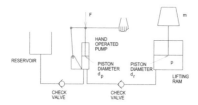

Hand operated hydraulic jack system

Characteristic of item	Size	Units
Hand pump piston diameter, d_p	0.25	in.
Hand pump piston stroke, ℓ_s	5.0	in.
Lifting ram piston diameter, d_r	2.75	in.
Mass being lifted, m	2000	lb$_\text{m}$
Height moved by weight, h	15.0	in.
Duration of lift phase, Δt	4.0	min.

3.5 A hydraulic motor is used to drive the rear wheels of a truck through a drive shaft. The truck is equipped with an accumulator that stores the kinetic energy of the truck during deceleration.

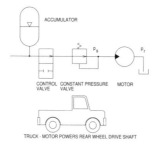

Determine the kinetic energy, KE, that can be stored in the accumulator if the truck is travelling at a velocity 45 mph and the energy is transferred to the accumulator with a pump as the truck decelerates to 0 mph. Determine the power, P, that is available to drive the truck if the energy from the previous part is used in a 30 s time interval. Determine the acceleration of the rear wheels caused by the pressure, p_s, applied to the motor. Determine the acceleration of the rear wheels caused by the pressure, p_s, applied to the motor when a load torque, T_L, of 1550 lbf · in. develops at the rear wheels. NOTE: T_L is the combined load torque for both rear wheels. Determine the acceleration of the rear wheels for the conditions given in the previous part , but with the mass of the truck, m, included in the calculation.

Simplified regenerative energy storage for a truck

Characteristic	Size	Units
System pressure, p_s	4500	$lb_f/in.^2$
Motor return pressure, p_r	50	$lb_f/in.^2$
Motor displacement, D_m	3.4	$in.^3/rev$
Rear wheel diameter, d	35.0	in.
Truck mass, m	3800	lb_m
Moment of inertia, I_c	177	$lb_f · in. · s^2$
(I_c is combined moment for		
2 rear wheels and drive shaft)		

3.6 A fluid power pump is used to drive a motor as shown in the figure. The flow velocity equals the line velocity thus $v_1 = v_2 = v_3 = v_4$.

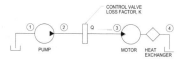

Determine the Reynolds Number, Re, fluid friction pressure loss, Δp_f, and the valve geometry loss, Δp_k, in the flow line between stations 2 and 3. Determine the power, P_m, produced by the motor. Determine the heat flow, Q_h J/h, that must be removed between stations 1 and 4 to maintain the final temperature, Θ_4.

Heat generation in a pump and motor transmission system

Characteristic	Size	Units
Pump power input, P_p	20	kW
Pump inlet pressure, p_1	0	
Pump outlet pressure, p_2	17.0	MPa
Pump inlet oil temperature, Θ_1	85.0	$°C$
Valve loss factor, K	200	
Total line length, ℓ_L	8.0	m
Line diameter, d	30.0	mm
Motor outlet pressure, p_4	0	Pa
Motor outlet temperature, Θ_4	85.0	$°C$
Oil specific heat, C_p	2002	$J/kg \cdot °\,C$
Oil density, ρ	850	kg/m^3
Oil viscosity, μ	9.4	$mPa \cdot s$

3.7 A fluid power pump is used to drive a cylinder as shown in the figure.

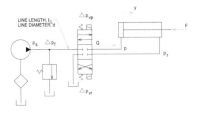

Determine the fluid friction pressure loss, Δp_f, and the valve losses, Δp_{vp} and Δp_{vr}. Determine the required pressure, p, at the piston to drive the cylinder. Determine the required pump speed, n rpm.

Characteristics of a pump and cylinder system

Characteristic	Size	Units
Cylinder piston diameter, d_p	42.0	mm
Cylinder rod diameter, d_r	20.0	mm
Cylinder load force, F	17000	N
Piston velocity, $\dot{y}$	0.35	m/s
Pump displacement, D_p	32.0	mL/rev
Pump volumetric efficiency, η_V	95	%
Line length, ℓ_L	4.0	m
Line diameter, d	12.0	mm
Flow coefficient into cylinder, C_{dp}	0.60	
Flow coefficient for return, C_{dr}	0.62	
Valve flow area into cylinder, A_{vp}	15.0	mm^2
Valve flow area for return, A_{vr}	17.0	mm^2
Oil mass density, ρ	850	kg/m^3
Oil viscosity, μ	9.4	mPa $\cdot$ s

REFERENCES

1. Merritt, H. E. , 1967, *Hydraulic Control Systems*, John Wiley & Sons, New York, NY.

2. Paul-Munroe Rucker, Inc., 1994, *Fluid Power Designers' Lightning Reference Handbook*, 8th ed., Paul-Munroe Rucker, Inc., Whittier, CA.

3. Munson, B. R., Young, D. F., and Okiishi, T. H., 1994, *Fundamentals of Fluid Mechanics*, 2nd ed., John Wiley & Sons, New York, NY.

4. Blackburn, J. F., Reethof, G., and Shearer, J. L., 1960, *Fluid Power Control*, The M.I.T. Press, Cambridge, MA.

5. ASTM International, 1998, "Standard Viscosity-Temperature Charts for Liquid Petroleum Products", D341-93, West Conshohocken, PA.

6. Yeaple, F., 1995, *Fluid Power Design Handbook*, 3rd ed., Marcel Dekker, Inc., New York, NY.

7. Keller, G. R., 1985, *Hydraulic System Analysis*, Hydraulics & Pneumatics Magazine, Penton/IPC, Cleveland, OH.

8. Akers, A., 1973, "Discharge Coefficients for an Annular Orifice with a Moving Wall", *Proceedings Third International Fluid Power Symposium*, BHRA, Turin, Italy, pp. B3-B37 and B3-B52.

9. Esposito, A., 1999, *Fluid Power with Applications*, 5th ed., Regents/Prentice Hall, Englewood Cliffs, NJ.

10. Fitch, E. C., 1968, *Hydraulic Component Modeling Manual*, Oklahoma State University, Stillwater, OK.

4

DYNAMIC MODELING

4.1 DEVELOPMENT OF ANALYTICAL METHODS

Fluid power systems can be modeled with mathematical equations such as those described in Chapters 2 and 3. As noted in those chapters, a variety of useful equations can be derived from basic physical principles derived from thermodynamics and fluid mechanics. In this chapter, dynamic effects resulting from accounting for Newton's Second Law and fluid bulk modulus will be introduced. These principles and a few concepts from vector mathematics, will provide a complete model for most fluid power systems.

Many very good fluid power machines have been developed with the use of functioning prototypes and trial and error testing. Operational evaluation of fluid power machines will always be desirable to provide proper operating conditions and safety. Early attempts to model systems with equations were mainly used to assist the testing of actual hardware. Modern modeling methods, however, now actually allow design and preliminary evaluation of fluid power machines before they are assembled. This change occurred as modeling methods became better understood and electronic computers became readily available.

Development of good modeling methods has resulted from the comparison of analytical results to laboratory test data. Many individuals have contributed to this work. One of the most notable published works, which is still very useful, is the result of a group effort [1]. This pioneer publication showed that mathematical models can be written that will be good representations of fluid power systems.

The equations that result when a fluid power system is modeled are difficult to solve analytically. The equations typically include first and second order differential equations, transcendental equations, and a variety of algebraic expressions. The operation of the system being modeled generally imposes several discontinuities on the equations. Early attempts to solve equation sets usually involved the use of analog computers or FORTRAN programming. Many approaches for model solution are now available. No attempt to list the methods in use would remain current for more than a few months. Fluid power periodicals provide a frequent list of methods with current features [2, 3]. The available software can generally be classified with regard to features. It must be noted, however, that features often overlap as each type of software undergoes programming improvements. The sections that follow provide a brief overview of solution methods and their abilities.

4.2   SOFTWARE OPTIONS

Several types of software have been written to facilitate arrangement and solution of equations. Some types of software, however, do not require the use of descriptive equations explicitly. Computers have become more powerful and graphical user interfaces are now standard. As a result there has been a progression from programming equations in a specific language to writing the equations in a conventional mathematical form and finally to drawing circuits with dedicated symbols for specific components.

The authors of this text, however, believe that some familiarity with equation development and solution is desirable for a fluid power engineer. Dedicated computer programs for fluid power are undoubtedly essential for an engineer who designs equipment on a daily basis, but a newcomer will benefit from starting from fundamentals. Expressing a physical system in a mathematical form requires simplifying assumptions and a knowledge of potential solution difficulties.

4.2.1   Equation Solutions

Equations written to describe fluid power machines may be solved with any of the common computer languages that handle mathematical statements. The descriptive equations can be arranged in the appropriate code and processed resulting in numerical or plotted results.

Equations may also be solved with symbolic computer software. This approach does not require that the equations be converted into a computer language. The descriptive statements can be entered as mathematical equations and the program will convert these equations into code suitable for execution without further effort by the user.

Use of direct equation solutions allows the user to have complete control of the machine modeling process. All of the necessary steps are clearly in view and easily revised as desired. Since the modeling equations are complex and sometimes require the use of assumptions, ready access to equations is a major advantage.

4.2.2 Graphical Solutions

Some types of software allow the arrangement of an engineering system in a graphical format. Mathematical equations are not required to use this type of software. The physical operation of the machine is described by the graphical arrangement. Elements and components of the system are selected from a menu and connected as desired. Specific information for each item is then entered in an appropriate table. The software recognizes the type of system being modeled and accounts for many characteristics such as discontinuities. Solution takes place directly from the graphical representation.

This type of software is very useful and is being used on many applications. Users must be careful to understand the physical link between the graphical model and the actual machine being modeled.

4.2.3 Fluid Power Graphical Symbol Solutions

Oil hydraulic machines are usually described with a system of fluid power graphical symbols, which are in worldwide use. They are the International Standard ISO 1219 fluid power system and component-graphical symbols [4]. Because these symbols are in common use, several types of software have been written that allow system solution from the graphical schematic. This type of schematic does not describe the physical operation of the machine. The software, however, provides for the description of each item in the system with a table of information. The software then recognizes the type of system being modeled and accounts for its specific characteristics. Solution takes place directly from the graphical schematic and the tabular information entered.

4.3 DYNAMIC EFFECTS

Chapter 3 presented some examples of systems that were truly time invariant and those that varied with time, but dynamic aspects could be ignored without much affecting the validity of the solution. As components move faster or as components become very massive, rates of change in pressure caused by fluid compliance and forces resulting from mass acceleration may become significant. We shall now incorporate the compliance and acceleration effects in system models. The formulation of such equations will be encountered often during the remainder of this text.

4.3.1   Fluid Compliance

The fluid property of bulk modulus and the related topic of container compliance was introduced in Chapter 2. Equation 2.5, $\beta = -\Delta p / (\Delta V / V)$ was presented. This equation can now be rearranged as the first building block in the development of dynamic equations.

$$\frac{dp}{dt} = \frac{\beta}{V} \frac{dV}{dt} \tag{4.1}$$

The minus sign has been removed because the formulation of dV/dt that is conventional results in a positive value of dp/dt.

Strictly, we should consider conservation of mass. In practice it is commonly assumed that the fluid density does not change much with pressure and a conservation of volume approach may be used with sufficient accuracy for most fluid power analyses. Although a conservation of volume approach is used for the flow through the passages, the effect of compressibility is accounted for when considering bulk volumes of fluid undergoing pressure change.

In fluid power circuits, there are volumes of fluid that exist between components. For example, consider the volume of fluid between a valve orifice and the piston of an actuator. There will be a flow of fluid into this volume and a flow out. The fact that the fluid volume may change with time is accounted for by making V vary with time. At any instant of time, the rate at which fluid in the volume V is being *compressed* (i.e. $dV/dt < 0$, hence the sign adjustment discussed previously) is given by:

$$\frac{dV}{dt} = \left(\sum_{i=1}^{i=m} Q_{in} - \sum_{j=1}^{j=n} Q_{out} \right)$$

The walls of the volume are considered totally rigid and any container compliance is accounted for by adjusting the effective bulk modulus as discussed in Chapter 2. This volume change can be expressed as a rate of *increase* in pressure by using Equation 4.1:

$$\frac{dp}{dt} = \left(\frac{\beta}{V(t)} \right) \left(\sum_{i=1}^{i=m} Q_{in} - \sum_{j=1}^{j=n} Q_{out} \right) \tag{4.2}$$

Notice that the fact that V may vary with time has been accommodated by writing the fluid volume as $V(t)$. The inflows, typically Q_{in}, and the outflows, typically Q_{out}, may be true flows such as that through an orifice or they may be caused by movement of the volume wall. For example, $Q = A_p v_a$, that is piston area times piston velocity. As with any mathematical formulation, the analyst must examine signs carefully. Equation 4.2 is written so $dp/dt > 0$ when the flow *into* the fluid volume exceeds the flow *out*. It should be recognized that the terminology compliance is simply the application of continuity discussed in Chapter 3

4.3.2 Newton's Second Law Effects

The second building block in the formulation of dynamic equations results from noting that masses subjected to unbalanced forces must accelerate according to Newton's Second Law, $F = ma$. This equation is commonly rearranged as:

$$a = \ddot{x} = \left(\sum_{i=1}^{i=n} F_i \right) \bigg/ m$$

As before, careful attention to signs is necessary. A direction for x must be established and the signs of forces specified accordingly. An allusion to vector concepts was made in the introduction to this chapter. It will be obvious that it is only components of forces resolved in the x direction that are included in the equation. An exception to this statement might be friction forces that may result from forces perpendicular to the direction of motion.

The previous equation is not suitable for most standard differential equation solvers. This second order equation must be split into two first order equations:

$$\frac{dx}{dt} = v \tag{4.3}$$

$$\frac{dv}{dt} = \left(\sum_{i=1}^{i=n} F_i \right) \bigg/ m \tag{4.4}$$

The forces on the moving parts of a fluid power circuit shown in Equation 4.4 may result from pressure differences, viscous friction, Coulomb friction, gravity, and flow forces to name a few.

4.4   WORKED EXAMPLES

Fortunately, most useful fluid power circuits can be simulated by solving the differential equations numerically. Two simple circuits will be presented.

In the first a vertical actuator supporting a mass load is controlled by a servovalve. A servovalve is one in which careful manufacture has ensured that the deadband at the null position is so small that it can be ignored. The ports are considered to be rectangular when projected on the cylindrical surface of the valve spool. The circumferential width, the area gradient, is w. Thus the open orifice area will be proportional to displacement. It will be observed that both up and down motion can be controlled by the valve. Consequently the fluid loses energy passing through two orifices. This is a penalty that must be paid for an actuator where bidirectional control is required.

As indicated earlier, dynamic modeling is required when rapid motion is required or when loads are massive. The required input to the valve has been chosen so that rapid motion is achieved.

In the second example, feedback has been introduced so the operating lever moves over the same displacement as the actuator. This example builds on the first example and the equations are very similar.

4.4.1   Example: Actuator Controlled by a Servovalve

A schematic of a servovalve controlled actuator is shown in Figure 4.1. It will be assumed that the servovalve is controlled by an actuator for which the dynamics may be ignored. Thus the only part of the circuit for which $F = ma$ must be written is for the actuator and its load. Observe that this actuator is considered to be a unit with equal areas at both sides ($A_{cap} = A_{rod}$).

The primary force causing motion is the pressure difference across the piston. This force is resisted by the viscous drag of the seal. For display purposes, this drag will be based on viscous drag presented in Equation 2.2. It will be seen that the drag force is directly related to velocity. The equation is repeated here using different symbols:

$$F_{sl} = \frac{\pi d_a \ell_{sl} \mu}{\ell_{gap}} v_a = c v_a$$

or:

$$c = \frac{\pi d_a \ell_{sl} \mu}{\ell_{gap}}$$

This form of drag force may be quite suitable for a valve spool in a bore, but it is less suitable in this context. Actuators are one of the few fluid power components that use rubbing, elastomeric seals. This is done to control fluid loss to the environment. The use of simple viscous drag in this example is for the sake of reducing equation complexity. More realistic seal drag forces will be presented in a later section (Section 4.5). Using the values presented in Table 4.1:

$$c = \frac{\pi \times 0.05 \text{ m} \times 0.01 \text{ m} \times 14.3\text{E}{-}3 \text{ Pa} \cdot \text{s}}{0.02\text{E}{-}03 \text{ m}} = 1.17 \text{ N} \cdot \text{s/m}$$

The other force on the actuator is the gravity force acting on the mass, $m_a g$. Given the chosen positive direction for x_a shown in Figure 4.1 this force will be shown as negative in the $F = ma$ equation.

So for the conditions presented, the Newton's Second Law equations are:

$$\frac{dx_a}{dt} = v_a \tag{4.5}$$

$$\frac{dv_a}{dt} = (p_1 A_{cap} - p_2 A_{rod} - c V_a - m_a g)/m_a \tag{4.6}$$

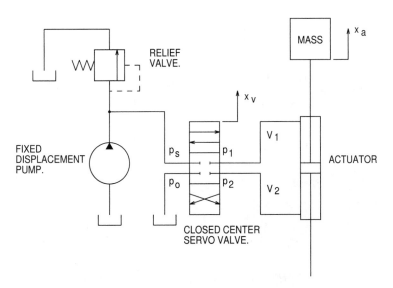

Figure 4.1: Servovalve controlled actuator.

Table 4.1: Characteristics of a servovalve controlled actuator

Characteristic	Value	Units
Mass, m_a	750	kg
Actuator piston diameter, d_a	0.05	m
Actuator rod diameter, d_{rod}	0.025	m
Oil volume, valve to port 1, V_1	15.7E−6	m^3
Oil volume, valve to port 2, V_1	15.7E−6	m^3
Maximum oil length in actuator, ℓ_a	0.5	m
Piston length, ℓ_{sl}	0.01	m
Piston annular gap, ℓ_{gap}	20.0E−6	m
Valve area gradient, w	0.01	m
Maximum valve opening, x_{vmax}	0.5E−3	m
Orifice discharge coefficient, C_d	0.62	
Acceleration of gravity, g	9.81	m/s^2
Fluid density, ρ	855	kg/m^3
Fluid bulk modulus, β_e	1.2E+9	Pa
Fluid absolute viscosity, μ	15.0E−3	Pa · s
System pressure, p_s	20.0E+6	Pa
Drain pressure, p_o	0.0E+6	Pa
Ramp up time, t_1	0.05	s
Begin ramp down time, t_2	0.95	s
Total valve open time, t_3	1.0	s

In order to simplify the appearance of the compliance equations somewhat, the flows through the valve orifices will be introduced as Q_1 and Q_2:

$$Q_1 = C_d w x_v(t) \sqrt{\frac{2(p_s - p_1)}{\rho}} \quad \text{and} \quad Q_2 = C_d w x_v(t) \sqrt{\frac{2(p_2 - p_o)}{\rho}}$$

Table 4.2: Servovalve controlled actuator, state variable initial values

State variable	Value	Units
Actuator displacement, x_a	0	m
Actuator velocity, v_a	0	m/s
Side 1 pressure, p_1	5.0E+6	Pa
Side 2 pressure, p_2	0	Pa

thus the equations will be:

$$\frac{dp_1}{dt} = \left(\frac{\beta_e}{V_1 + A_{cap}x_a}\right)(Q_1 - A_{cap}v_a) \tag{4.7}$$

$$\frac{dp_2}{dt} = \left(\frac{\beta_e}{V_2 + A_{rod}(l_a - x_a)}\right)(A_{rod}v_a - Q_2) \tag{4.8}$$

Thus this system requires 4 ordinary differential equations to describe its operation. The expression *state variable equations* may also be encountered. A state variable is simply a quantity that appears as a first derivative on the left-hand side of an ordinary differential equation. In this example, x_a, v_a, p_1, and p_2 are state variables. It should also be recognized that most state variable equations that are likely to be encountered in the analysis of fluid power systems will not have analytical solutions. As indicated earlier in this chapter, the general availability of computers in engineering has meant that virtually any set of ordinary differential equations that may be formulated for a fluid power system can be solved numerically.

If a designer is solving differential equations numerically, the program will generally require that the user provide initial values of the state variables at the beginning of the integration interval. Sometimes all zeros are suitable values, but in this instance it is obvious that the weight of the load, $m_a g$ must be supported by the initial value of the pressure p_1. This has the value:

$$p_1 = \frac{m_a g}{\pi(d_a^2 - d_{rod}^2)/4} = \frac{750 \text{ kg} \times 9.81 \text{ m/s}^2}{\pi((0.05 \text{ m})^2 - (0.025 \text{ m})^2)/4} = 5.0\text{E+6 Pa}$$

No real valve can be opened instantly so the valve input function has been presented as a ramp up to maximum value in t_1, a steady period of constant valve area from t_1 to t_2, and a closing from t_2 to t_3.

Some results of the simulation when $t_3 = 1.0$ s are shown in Figures 4.2 to 4.4 Once the valve is held open with a constant area, the flow will

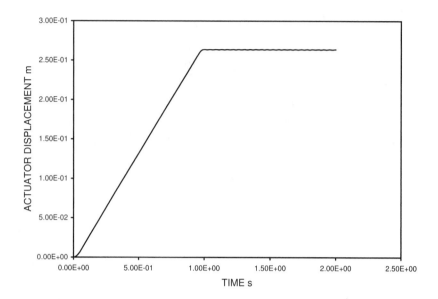

Figure 4.2: Servovalve controlled actuator, displacement vs. time.

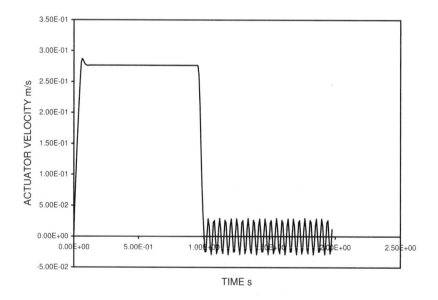

Figure 4.3: Servovalve controlled actuator, velocity vs. time.

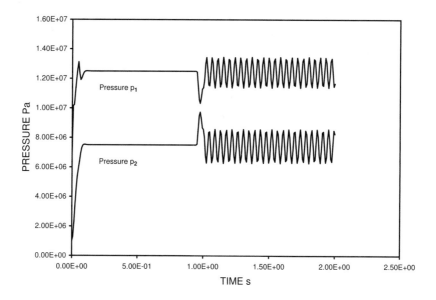

Figure 4.4: Servovalve controlled actuator, pressures vs. time.

become constant after the decay of transients. This is shown both in the displacement and in the velocity figures. The displacement is mainly a linear increase and the velocity becomes constant. The velocity figure shows clear dynamic effects. After the valve has closed and there is no damping effect of the fluid passing through the orifice, the moving mass is supported on the oil spring described in Chapter 2. The oscillations in velocity are quite clear. There are accompanying oscillations in displacement, but these are small and are more easily seen by inspecting the tabular output from the simulation.

The dynamic analysis for pressure (Figure 4.4) is even more different from the results that would be expected from a pseudo steady state simulation. The mass requires force to accelerate it from rest so the pressure p_1 shows a some initial overshoot. On the other hand, when the valve closes and the mass is still moving, it must be brought to rest again. The pressure p_1 falls initially, but then oscillates about 12.5E+6 Pa. Likewise the pressure p_2 rises initially and then oscillates about 7.5E+6 Pa.

The difference between the mean values of p_1 and p_2 is 5E+6 Pa as would be expected from the static calculation performed to determine a suitable initial value for p_1. The increases in both p_1 and p_2 from their postulated initial values can be explained by the fact that these pressures were achieved during the steady state motion and they are locked in when the valve closes.

The frequency of oscillations observed in this simulation can be compared with the frequency (Equation 5.25) that would be predicted by using the equation developed for the stiffness of an oil spring (Equation 2.16).

$$k = \beta_e \left(\frac{A_1^2}{V_1} + \frac{A_2^2}{V_2} \right) \tag{2.16}$$

Noting that the lower oil column length is $x_a = 0.264$ m and the upper $l_a - x_a = 0.5 - 0.264 = 0.236$ m and adding in the oil volumes in the supply passages from the valve:

$$
\begin{aligned}
k &= 1.2\text{E}+9 \, \frac{\text{N}}{\text{m}^2} \times \\
&\quad \left(\frac{(0.001473 \text{ m}^2)^2}{(388.5\text{E}-6 + 15.7\text{E}-6) \text{ m}^3} + \frac{(0.001473 \text{ m}^2)^2}{(347.8\text{E}-6 + 15.7\text{E}-6) \text{ m}^3} \right) \\
&= 13.6\text{E}+6 \text{ N/m}
\end{aligned}
$$

Thus the expected frequency will be given by:

$$\omega_n = \sqrt{k/m_a} = \sqrt{\frac{13.6\text{E}+6 \text{ N/m}}{750 \text{ kg}}} = 135 \text{ rad/s}$$

Extracting values from the simulation tabular output yielded an oscillation period of 0.048 s, i.e. a frequency of 130 rad/s. The analysis just performed is not exact. The two volumes of oil in the passages from the valve to the actuator should be treated as separate spring rates because their geometry is different from the oil in the actuator. The result obtained, however, is sufficiently close to the simulation results to give considerable credibility to the simulation.

The seemingly steady oscillation amplitude of the velocity and pressures after the valve closes can be explained by examining the value of damping coefficient for the system. Using Equation 5.24:

$$
\begin{aligned}
\zeta &= \frac{1}{2} c \sqrt{\frac{1}{k m_a}} \\
&= \frac{1}{2} 1.17 \text{ N} \cdot \text{s/m} \sqrt{\frac{1}{13.6\text{E}+6 \text{ N/m} \times 750 \text{ kg}}} \\
&= 5.79\text{E}-6 \text{ (dimensionless)}
\end{aligned}
$$

This is a very low damping coefficient and explains why no obvious attenuation of oscillation is observed on Figures 4.3 and 4.4. The low value of damping coefficient is explained by the low value of viscous damping, c, and the high values of spring rate, k, and mass, m_a. This is a good example of why simulation is important. Recognizing that the damping coefficient, ζ, is very low might suggest to the designer that examining the damping magnitude might be necessary. The role of Coulomb friction for rubbing elastomeric seals is discussed in Section 4.5.

4.4.2 Example: Hydromechanical Servo

Two of the major attractions of fluid power are that force can be magnified and that feedback can be provided. Thus a movement can be given to a control lever by an operator and a device will mimic the movement of the lever. That is, the device will stop when the control lever is stopped. Compare this to an actuator controlled by a proportional valve as examined in the previous section. If the operator wants to stop the actuator, then he or she must return the valve to a centered position. Obviously this is a perfectly satisfactory situation in many instances, for example, consider a backhoe operator performing a digging operation. Working the boom and the bucket with proportional control valves that must be centered to stop motion serves quite adequately.

On the other hand, the same operator would probably not be satisfied by a steering system that had to be returned to a central position once the wheels were brought to point in the desired position. Here a feedback system is needed that points the vehicle's wheels in the direction of the steering wheel and holds them there until the steering wheel is moved again. In virtually all small passenger vehicles, power steering is totally mechanical and will use a similar system to that shown in Figure 4.5.

Consider the lever AC starting in a horizontal position. The actuator would be in the center of its travel and the directional control valve would also be centered, so there would be no flow to the actuator. Incidentally, notice that the actuator is symmetrical. That is, the piston area is the same for both sides. Now let point A on the lever AC be moved some distance up (the input direction on Figure 4.5). Initially point D will be stationary so point E moves down. The valve now opens and fluid flows into the lower half of the actuator and out of the upper half. Thus point D moves up and in so doing moves point E up also. The spool will move up until the valve is exactly centered and flow to the actuator ceases. Notice that the feedback is negative, which is required for a stable automatic control system. The system is now in a state of equilibrium again, but the lever end A has moved up, as has the output end of the actuator. This mode

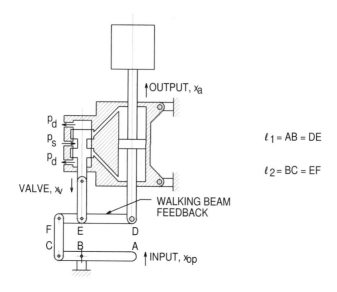

Figure 4.5: Walking beam feedback for servovalve controlled actuator.

of operation is followed as the lever AC is moved to other positions. Lever DF provides feedback between the actuator (output) and the input (the directional control valve). In fact, lever AC was provided so output was in the same direction as input. This is usually the desired situation.

The analysis of this system with feedback (closed loop) is very closely related to the analysis of the actuator without feedback (open loop). Consider the dimensions:

$$\ell_1 = AB = DE$$
$$\ell_2 = BC = EF$$

Initially for point D stationary, the valve movement for operating lever movement (point A) x_{op} is:

$$x_v = \frac{\ell_2}{\ell_1}\frac{\ell_1}{\ell_1 + \ell_2}x_{op} = \frac{\ell_2}{\ell_1 + \ell_2}x_{op}$$

Now consider A stationary and some movement of the actuator, x_a. The resulting valve movement will be:

$$x_v = -\frac{\ell_2}{\ell_1 + \ell_2} x_a$$

Thus the relationship between x_v, x_{op}, and x_a is:

$$x_v = \frac{\ell_2}{\ell_1 + \ell_2}(x_{op} - x_a) \tag{4.9}$$

All that needs to be done to simulate this closed loop system is to insert Equation 4.9 into the open loop simulation and make the input function $x_{op}(t)$ instead of $x_v(t)$. The value of x_a will be updated throughout the simulation because x_a is a state variable.

The simulation of this system was performed with the characteristics presented in Table 4.3. The forcing function was a modified square wave where the rise or fall was ramped and not instantaneous. For simplicity in programming, the driver function was generated by truncating a larger amplitude sine wave. Two periods were examined, 2.0 s and 10.0 s. The results for displacement and the lower and upper pressures in the actuator are presented in Figures 4.6 to 4.9. The simulation results using the short period driver, 2.0 s, are not very satisfactory. The frequency, $\omega = 3.14$ rad/s or f = 0.5 Hz, is quite high for a system moving a mass of 750 kg over a distance of ± 0.1 m, The output of the actuator does not track the operator input closely. As might be expected, the pressure results show pronounced spikes whenever there is a change in direction of the actuator.

On the other hand, the results for the long period driver, 10.0 s, are much more satisfactory. The actuator output is able to track the input reasonably closely. The pressure fluctuations are much less evident, as would be expected from the lower rates of change of actuator velocity.

Although the hydromechanical servo shown in Figure 4.5 is a satisfactory simple example of a servo system with mechanical feedback, a real life system would require additional features. For example, there is no limit on the size of actuator shown or the pressure available at p_s. In theory this would mean that the operator of a heavy truck could steer the truck with one finger. In practice this is not a satisfactory approach and most servo systems would incorporate some degree of force feedback so the operator has a feeling about the magnitude of the task that is being performed. Another point, in the system discussed it would be impossible to steer if there was no fluid pressure. A typical car power steering system would have the steering wheel directly connected to the steering system to allow operation when the engine was turned off.

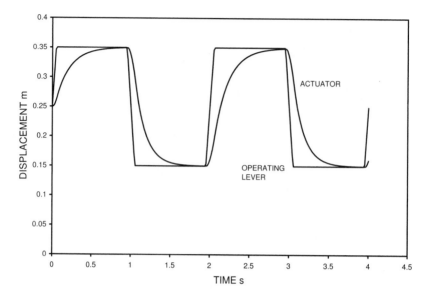

Figure 4.6: Hydromechanical servo, displacement vs. time for short period driver.

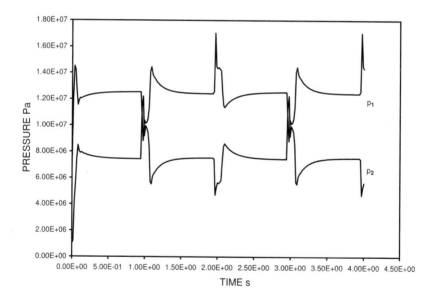

Figure 4.7: Hydromechanical servo, pressure vs. time for short period driver.

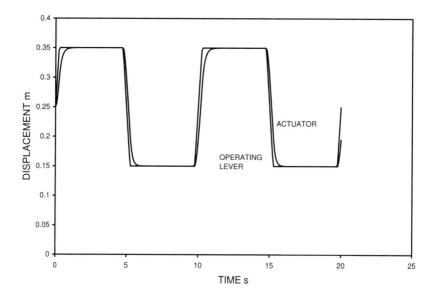

Figure 4.8: Hydromechanical servo, displacement vs. time for long
period driver.

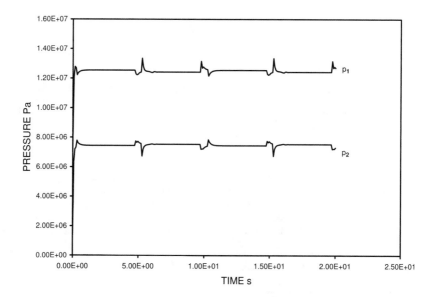

Figure 4.9: Hydromechanical servo, pressure vs. time for long
period driver.

Table 4.3: Characteristics of a hydromechanical servo system

Characteristic	Value	Units
Mass, m_a	750	kg
Actuator piston diameter, d_a	0.05	m
Actuator rod diameter, d_{rod}	0.025	m
Oil volume, valve to port 1, V_1	15.7E−6	m^3
Oil volume, valve to port 2, V_1	15.7E−6	m^3
Maximum oil length in actuator, ℓ_a	0.5	m
Piston seal length, ℓ_{sl}	0.001	m
Valve area gradient, w	0.01	m
Orifice discharge coefficient, C_d	0.62	
Acceleration of gravity, g	9.81	m/s^2
Fluid density, ρ	855	kg/m^3
Fluid bulk modulus, β_e	1.2E+9	Pa
Fluid absolute viscosity, μ	15.0E−3	Pa · s
System pressure, p_s	20.0E+6	Pa
Drain pressure, p_o	0.0E+6	Pa
Driver period, t	2.0	s
or		
Driver period, t	10.0	s
Walking beam dimension, ℓ_1	1.0	m
Walking beam dimension, ℓ_2	0.01	m

4.5   MODELING HINTS AND TIPS

Pressure reversal in an orifice: The standard equation for an orifice, Equation 3.19, may give trouble to a designer new to modeling.

$$Q = C_d A \sqrt{\frac{2\Delta p}{\rho}}$$
(3.19)

The problem occurs because the term Δp may reverse its sign. Computer programming languages object to finding the square root of a negative number. The problem is easily solved by introducing a multiplying factor on the right-hand side of the coded equation. This factor, here called SGNP, will be -1 if Δp is negative, 0 if Δp is zero, and $+1$ if Δp is positive. Now the flow direction has been associated with SGNP, Δp is replaced by $|\Delta p|$ in the coded equation.

Treatment of rigid stops: Many valves have an element or spool that moves under the influence of a pressure difference. Typically the element is brought to rest at the ends of its travel by a rigid stop. Although the effect of such stops may be modeled by some IF THEN ELSE statement that makes the velocity zero when the spool tries to make an extreme excursion, such an approach may render the equation difficult to integrate. The problem can often be solved by replacing the stop by a very rigid spring that only affects the dynamics for attempted extreme spool excursions. If the spring rate is chosen judiciously, the extreme excursion beyond the stop will be very small, yet the system dynamics can be handled by the differential equation solver. The analyst must also decide how to handle a negative valve opening that may occur with this strategy. Usually an IF THEN ELSE statement setting negative areas to zero will suffice without affecting integration.

Flow reversal through a four-way valve: In the example of a hydromechanical servo discussed in Section 4.4.2, the valve legitimately made negative and positive excursions about a central null position. This is different situation from that discussed the previous paragraph. The code must account for the fact that the supply and drain pressures are switched between the ports. This can easily be done with IF THEN ELSE statements based on the sign of the valve excursion. Another factor that must be introduced into the code is that the areas that appear in Equation 3.19 must appear as absolute values. Finally inspect the equations derived from Equation 4.2, redisplayed here, to ensure that they behave as expected with the pressure switching.

$$\frac{dp}{dt} = \left(\frac{\beta}{V(t)}\right)\left(\sum_{i=1}^{i=m} Q_i - \sum_{j=1}^{j=n} Q_o\right) \tag{4.2}$$

Valve deadband: Because numerical solutions are more powerful than analytical solutions, the iterative solution using a differential equation solver can also handle deadband. Making a valve with a vanishingly small deadband, as was postulated in Section 4.4.2, is expensive and may not be justified. Deadband can be handled by a suitable IF THEN ELSE statement that has zero valve opening over a small range of valve travel.

Handling negative pressures: There may be situations when oscillatory behavior is being simulated when large negative pressure excursions are predicted. This problem can usually be handled by an IF THEN ELSE statement limiting negative gauge pressures to $-1.0\mathrm{E}{+}5$ Pa (i.e. negative standard atmospheric pressure). Although this is a recognition that the absolute pressure cannot fall below zero, it is also an indication that cavitation will occur. The design or operating conditions should be reappraised to eliminate the cavitation.

Coulomb friction and actuator seals: Actuator seals differ from most internal seals found in fluid power equipment. Actuators are often required to hold their position when a valve has been returned to a null position so leakage must be minimal. Consequently, actuators are usually fitted with tight fitting elastomeric seals that are further enhanced by letting pressure force the seal lip against the cylinder bore or the rod. Thus actuator seals usually exhibit Coulomb and not viscous friction characteristics. Many combinations of solid materials exhibit a higher coefficient of friction before motion commences than afterwards. A simple equation that can represent this situation is:

$$\mu_{st}(|v|) = \mu_{stmax}\left(\frac{1}{(c|v|+b)^n}+1\right) \tag{4.10}$$

This equation has been plotted in Figure 4.10. The four parameters, b, c, n, and μ_{stmax} should be determined experimentally for the specific seal and metal combination. Elaborate curve fitting is not required. Once μ_{stmax} and the running value of μ_{st} have been determined:

$$b = \frac{\mu_{stmax}}{\mu_{strun}} - 1$$

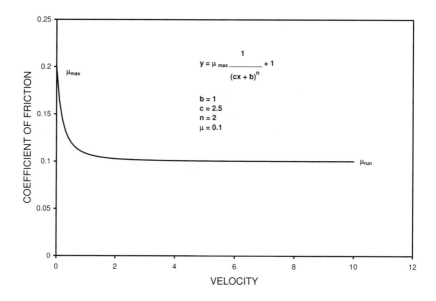

Figure 4.10: Simplified curve for simulating Coulomb friction.

and c and n can be set with a few trials until an acceptable fit is obtained. The coefficient c is intended to account for the units used for velocity and n determines how rapidly the coefficient of friction drops after motion commences. Once the expression for the coefficient of friction has been established, it is used to predict the drag force generated by the seal from:

$$F_{seal} = SGNV p A_{seal} \mu_{st}(|v|) \qquad (4.11)$$

Where SGNV is -1 for negative velocity, 0 for zero velocity, and $+1$ for positive velocity.

Differential equation solvers: A differential equation set that results from analyzing a fluid power system, may contain displacements, velocities, or pressures that change very rapidly and state variables that change much more slowly. That is, there will be a very wide range of time scales present in the solution to the equation set. Such equations are called *stiff* in mathematics texts [5] and these equations may need special techniques for integration.

The authors of this text have taught many undergraduate engineering students their first course in fluid power. Regrettably, this is usually their last course in fluid power also! It is our belief that newcomers to fluid power analysis will benefit from expressing devices and systems in the form of equations initially. These equations can be solved using a variety of software. Simulation of devices and systems almost always implies presentation of results in the form of graphs.

A program that deserves attention by the newcomer is Microsoft Excel®. This program offers a very comprehensive combination of graphing capability and a mathematical programming language, Visual Basic for Applications®. The example presented in Section 4.4.2 was solved in Excel® with the Kaps and Rentrop modification of the Rosenbrock method. This is an adaptive step size code for stiff ODEs. The code was translated from code in C++ [6].

Specialized simulation programs have an edge over using a user-written routine. Programs like Mathcad® and Matlab® have several options for solving differential equations. If a problem presents difficulties with a default, non stiff solver, then the user should try selecting a stiff method such as Gear's method.

4.6 DISCUSSION

The spring-mass-damper analytical model, the open loop valve controlled actuator, and the closed loop hydromechanical servo all demonstrate that dynamic modeling is necessary to account for oscillatory behavior. Dynamic modeling may be very useful when a compromise must be made between acceptable overshoot and time of response. As a general rule, a faster system is likely to have more overshoot and oscillation. Dynamic modeling will also alert the designer to possible harmful pressure transients.

There are two basic reasons for writing a mathematical model for a fluid power system. The primary benefit derived from establishing a model is to promote understanding in regard to system function. The model then allows evaluation of system operation, which provides the important second benefit. Computer software, whether specifically for fluid power analysis or general purpose, allows easy changes in system variables. Therefore, establishment of a mathematical model allows study of the system for the complete range of expected characteristic variation.

Information on modeling is available from many sources. Most of the texts and handbooks specifically written on fluid power systems will provide useful information. Computer software specifically written for fluid power analysis usually contains helpful information. Numerous publications are available that give detailed information for fluid power system modeling. Some typical examples are listed in the references at the end of this chapter

[7-11].

Mathematical expressions for fluid power systems are not exact, however, they will generally provide much useful performance information and will define the interaction among the selected system variables. Most models allow considerable flexibility with regard to what equations and characteristics are included. The model will have limitations on its use and applicability and must be used accordingly. Results gained from mathematical modeling must always be carefully examined with regard to expected system behavior. It then becomes possible to select those system characteristics or variables that have the greatest influence on system operation. A good model will ultimately lead to a better understanding of the system being studied. Also, it will define the magnitude of various characteristics to provide the desired system performance. The flexibility and overall usefulness of fluid power are very advantageous. These factors require careful consideration in regard to what is included in a model.

Available data and knowledge on a particular system will define the nature of the mathematical model. The establishment of a model will always lead to better understanding of system performance. Therefore, the time spent to write the model will be well rewarded.

PROBLEMS

4.1 Consider the problem given in Section 4.4.2. Simulate this problem in your choice of analysis program. Lower the value of the bulk modulus to represent air entrainment and the effect of flexible hoses until undesirable oscillations become apparent when the actuator makes the transition from up to down (or the reverse).

4.2 Mechanical devices exhibit inertia, so forcing a system to operate at higher and higher frequency will mean that the performance deteriorates as the frequency increases. Consider the problem given in Section 4.4.2. Start with a sinusoidal input without truncation. Examine periods of 31.6, 10, 3.16, 1, and 0.316 s. Comment on the changes observed for displacement, velocity, and pressure.

4.3 Consider the problem given in Section 4.4.2. The performance of the unit when the period is 2 s is poor. The actuator does not conform well to the input driver variation.

4.3.a Try altering the pressure to 30 Mpa. Comment on the changes observed for displacment, velocity, and pressure. Comment on any simulation result that might not be desirable.

4.3.b Return the pressure to 20 MPa and try doubling the value of the valve gradient. Comment on the changes observed for displacment, velocity, and pressure. Comment on any simulation result that might not be desirable.

REFERENCES

1. Blackburn, J. F., Reethof, G., and Shearer, J. L., 1960, *Fluid Power Control*, The M.I.T. Press, Cambridge, MA.

2. Doe, J., Current and earlier years, "Miscellaneous Articles on Modeling Programs", Fluid Power Journal, Innovative Designs & Publishing, Bethlehem, PA.

3. Doe, J., Current and earlier years, "Miscellaneous Articles on Modeling Programs", Hydraulics & Pneumatics, Penton Publishing, Inc., Cleveland, OH.

4. Paul-Munroe Rucker, Inc., 1994, *Fluid Power Designers' Lightning Reference Handbook*, 8th ed., Paul-Munroe Rucker, Inc., Whittier, CA.

5. Press, W. H., Flannery, B. P., Teukolsky, S. A., Vetterling, W. T., 1986, *Numerical Recipes The Art of Scientific Computing*, Cambridge University Press, Cambridge, U.K.

6. Press, W. H., Flannery, B. P., Teukolsky, S. A., Vetterling, W. T., 1992, *Numerical Recipes The Art of Scientific Computing*, 2nd ed., Cambridge University Press, Cambridge, U.K.

7. Fitch, E. C., 1968, *Hydraulic Component Modeling Manual*, Oklahoma State University, Stillwater, OK.

8. McCloy, D. and Martin, H. R., 1980, *Control of Fluid Power: Analysis and Design*, 2nd ed., Halsted Press: a division of John Wiley & Sons, Chichester, U. K.

9. Palmberg, J., 1994, "Fluid Power System Modeling and Simulation Techniques Commonly Used in Europe", NFPA Visions 2000 NFPA Conference, NFPA Chicago, IL.

10. Richards, C. W., Tilley, D. G., Tomlinson, S. P., and Burrows, C. R., 1990, "Bathfp - A Second Generation Package for Fluid Power Systems", *Proceedings of BHRA 9th International Fluid Power Symposium*, BHRA, Cambridge, UK, pp. 315-322.

11. Watton, J., 1989, *Fluid Power Systems*, Prentice Hall, New York, NY.

5

LINEAR SYSTEMS ANALYSIS

5.1 INTRODUCTION

Thus far in the text, the goal has been simulation of devices or simple assemblies of devices. There are occasions when feedback is incorporated in a system and the magnitude of allowable feedback must be determined before a system is simulated. This is the subject of automatic controls.

It has been the experience of the authors that there are two serious challenges that may be faced by the primary audience of this text, senior undergraduate engineers. The first challenge is that a course on automatic controls seems to be very dense and of considerable mathematical complexity. The undergraduate may reach the end of a one semester course having been exposed to the many concepts of control theory without having been able to apply this theory to a physical problem to which they can really relate. The second challenge is that some engineering disciplines that use fluid power extensively are not able to find room for a controls course in the curriculum. For example, Agricultural Engineering is often in this class.

This chapter is an attempt to introduce the basic concepts of automatic control theory, so the fluid power engineer can relate individual devices to more complex systems incorporating feedback. This chapter will introduce the idea of a linear system. It will be shown how such systems may be analyzed by transformation methods. After transforming from the time domain to the s or Laplace domain, we shall demonstrate how physical entities can be laid out in *block* diagrams and how the block diagrams in the time domain can be converted into the Laplace domain. These block diagrams, however, are much more generic than fluid power circuits, yet considerable quantitative and qualitative insight may be gained from their analysis. These block diagrams will be manipulated so a network of simple blocks can be consolidated into a single function, the *transfer* function, that may be said to transform the input signal into the output. These consolidated functions will be used to determine the stability of a system.

Two fundamental entities, the spring-mass-damper and the single fluid volume with differential flow rates, are building blocks for fluid power systems. These two entities will be analyzed to introduce ideas concerning damping, resonant frequency, and a system time constant.

5.2 LINEAR SYSTEMS

The description *linear* has been encountered several times in this text. For dynamic systems, a linear system is one in which the equations of motion are *ordinary* differential equations with *constant* coefficients. A corollary of this statement is the fact that if two or more inputs are applied to a linear system *concurrently*, then the output will be the *sum* of the outputs that would occur if the inputs acted on the system individually.

At the level at which control theory will be introduced in this chapter, a system will either be linear or approximations will have been made so the system can be treated as linear. It should be stated categorically that approximating nonlinear systems by linear ones is not generally a valid approach for simulation. The linearized system can be investigated for stability properties near a specific operating point and this information can then be used in a more general nonlinear simulation.

The term *stability* will be explained in more analytical terms as the mathematical developments are continued. At this stage, consider a stable system as one that will have bounded response when subject to a bounded input. An unstable system is one in which the amplitude of the response will grow *exponentially* with time even for a bounded input.

5.3 THE LAPLACE TRANSFORM

Many engineering phenomena can be expressed in the form of ordinary differential equations. Although there have been iterative techniques for solving non linear ordinary differential equations for many years, the amount of calculation was essentially overwhelming for all but simple systems. The Laplace transform was a technique developed so certain differential equations could be solved for extended durations without needing huge numbers of calculations. With the advent of computers, Laplace transforms are not used for system simulation, but they can still give information on stability and are very useful for designing systems with feedback.

The Laplace transform is a process that converts a function in the time domain to a function in another domain. This other domain is the s domain and allows a wide variety of algebraic manipulations to be performed [1]. The formal definition is:

$$\mathscr{L}(f(t)) = L(s) = \int_0^\infty e^{-st} f(t) dt \qquad (5.1)$$

Decaying exponential: Consider the transform of a decaying exponential:

$$f(t) = e^{-at}$$

applying the transformation procedure:

$$
\begin{aligned}
\mathscr{L}(e^{-at}) &= \int_0^\infty e^{-st} e^{-at} dt = \int_0^\infty e^{-(s+a)t} dt \\
&= \frac{-1}{s+a} \left[e^{-(s+a)} \right]_0^\infty = \frac{-1}{s+a} \left[e^{-(s+a)\infty} - e^{-(s+a)0} \right] \\
&= \frac{1}{s+a} \qquad (5.2)
\end{aligned}
$$

Differential coefficients: Because the Laplace transform is associated with the solution of certain ordinary differential equations, the next example will investigate the transforms of differential coefficients. Consider the transform:

$$\mathscr{L}(f'(t)) = \int_0^\infty e^{-st} f'(t) dt$$

This can be approached in the following manner:

$$\frac{d}{dt}(e^{-st} f(t)) = -se^{-st} f(t) + e^{-st} f'(t)$$

Rewrite this as:

$$e^{-st} f'(t) = se^{-st} f(t) + \frac{d}{dt}(e^{-st} f(t))$$

Now apply the Laplace transform procedure:

$$
\begin{aligned}
\int_0^\infty e^{-st} f'(t) dt &= \int_0^\infty se^{-st} f(t) dt + \int_0^\infty \frac{d}{dt}(e^{-st} f(t)) dt \\
\mathscr{L}(f'(t)) &= sF(s) + \left[e^{-st} f(t) \right]_0^\infty \\
&= sF(s) - f(0^+) \qquad (5.3)
\end{aligned}
$$

The $f(0^+)$ notation indicates that $f(t)$ is evaluated at an infinitely small but positive value of t. This consideration is necessary because some functions of interest may be discontinuous at $t = 0$. This procedure can be extended to higher order differential coefficients, for example:

$$\mathscr{L}\left(\frac{d^2 f}{dt^2}\right) = s^2 F(s) - sf(0^+) - f'(0^+) \tag{5.4}$$

The standard practice in controls work is to consider that the system was at rest before any input was applied, thus:

$$f(0^+) = f'(0^+) = f''(0^+) = f'''(0^+) = \ldots f^n(0^+) = 0$$

consequently a differential equation of the form:

$$\frac{d^3 f}{dt^3} + a_1 \frac{d^2 f}{dt^2} + a_2 \frac{df}{dt} + a_3 f = g(t) \tag{5.5}$$

for controls purposes transforms to:

$$s^3 F(s) + a_1 s^2 F(s) + a_2 s F(s) + a_3 F(s) = G(s) \tag{5.6}$$

This expression may be rewritten:

$$\frac{F(s)}{G(s)} = \frac{1}{s^3 + a_1 s^2 + a_2 s + a_3} \tag{5.7}$$

Transfer function: It is now possible to introduce a little more terminology that is specific to the use of Laplace transforms in control theory. Rearrange Equation 5.7 in the form:

$$F(s) = \frac{1}{s^3 + a_1 s^2 + a_2 s + a_3} G(s) = T(s)G(s)$$

The inversion of $F(s)$ into the time domain will give the time response of the system when subject to the specific input $G(s)$ after it has been transformed by $T(s)$. As indicated earlier, this is not necessarily a good procedure for actually determining the output, but the formulation shows that $T(s)$ is a property of the system only. Thus $T(s)$ is called the *transfer function* of the system.

The integral: Another calculus related function is the integral. Integration blocks are very commonly encountered in fluid power or dynamic analysis where forces are applied to masses resulting in accelerations. One

integration of acceleration yields velocity and a second yields displacement.
Thus we need to evaluate:

$$\mathscr{L}\left(\int_a^t f(t)\right) = \int_0^\infty e^{-st}\int_a^t f(t)\,dt$$

This integral may be evaluated using integration by parts:

$$-\int u'v\,dt = \int uv'\,dt - uv$$

Where $u = e^{-st}/s$, $u' = -e^{-st}$, $v = \int_a^t f(t)$, and $v' = f(t)$ thus:

$$
\begin{aligned}
\int_0^\infty e^{-st}\int_a^0 f(t)\,dt &= \int_0^\infty \frac{e^{-st}}{s}f(t)\,dt - \left[\frac{e^{-st}}{s}\int_a^t f(t)\right]_0^\infty \\
&= \frac{1}{s}\mathscr{L}(f(t)) - \frac{1}{s}\left[0 - \int_a^0 f(t)\right] \\
&= \frac{1}{s}\mathscr{L}(f(t)) + \frac{1}{s}\int_a^0 f(t)\,dt
\end{aligned}
$$

As indicated earlier, in control theory it is normal to assume all zero conditions for $t \le 0$, thus the Laplace transform of an input can be taken as:

$$\mathscr{L}\left(\int_0^t f(t)\right) = \frac{1}{s}\mathscr{L}(f(t)) \tag{5.8}$$

Unit step: An input that starts at zero until $t = 0$ and then rises immediately to a value of one is the unit step function. Although such an input cannot be obtained in practice for real dynamic systems, the function is a useful conceptual input for examining the response of certain systems. The function is usually named $u(t)$ and is defined as:

$$\text{for } t < 0 \text{ then } u(t) = 0$$
$$\text{for } t \ge 0 \text{ then } u(t) = 1$$

The transform is:

$$\mathscr{L}(u(t)) = \int_0^\infty e^{-st}u(t) \tag{5.9}$$

$$= \left[\frac{e^{-st}}{-s}\right]_0^\infty \frac{-1}{s}[0-1]$$

$$= \frac{1}{s} \tag{5.10}$$

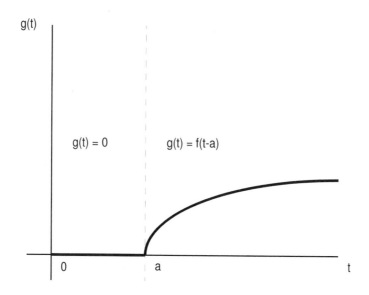

Figure 5.1: Second shifting theorem.

Second shifting theorem: Consider the function $g(t)$ shown in Figure 5.1. Apply the Laplace transform:

$$
\begin{aligned}
\mathcal{L}(g(t)) &= \int_0^\infty e^{-st} g(t) dt \\
&= \int_0^a e^{-st} 0 dt + \int_a^\infty e^{-st} f(t-a) dt \\
&= \int_a^\infty e^{-st} f(t-a) dt
\end{aligned}
$$

Now make the change in variables:

$$
\tau = t - a \ \text{ thus } \ d\tau = dt \ \text{ and } \ \tau = 0 \ \text{ when } \ t = a
$$

consequently:

$$
\mathcal{L}(g(t)) = \int_0^\infty e^{-s(\tau + a)} f(\tau) d\tau
$$

$$\begin{aligned} &= e^{-as} \int_0^\infty e^{-s\tau} f(\tau) d\tau \\ &= e^{-as} \mathscr{L}(f(\tau)) = e^{-as} \mathscr{L}(f(t)) \end{aligned} \tag{5.11}$$

Thus any function in the Laplace domain that is multiplied by e^{-as} will invert to a function in the time domain that is delayed by an amount $t = a$. The function has zero value until $t = a$.

Impulse: An impulse is defined mathematically as having an infinite amplitude for zero time. This might seems a rather useless concept. In fact the impulse can be approximated quite adequately in practice if the input duration is very short with respect to other time characteristics of the system such as time constants and periodic times. For an example, consider a bell. The bell can be made to ring by giving it a quick tap with a hammer. Most bells will be sufficiently massive that the hammer tap may be treated as a true mathematical impulse. The reason for exciting a system with a unit impulse or its approximation is that the transform of the impulse is just unity so the Laplace transform of the response of a system to an impulse is the transform of the system.

Consider the two step functions shown in Figure 5.2. Note that they

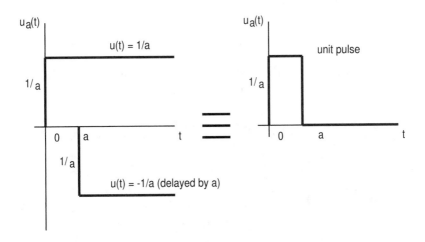

Figure 5.2: Pulse function.

are not unit step functions but have amplitude $1/a$. The combination of the two step functions yields a pulse of length a. We can use Equation 5.11 to write the Laplace transform of a pulse as:

$$\mathscr{L}(u_a(t)) = \frac{1/a}{s} - e^{-as}\frac{1/a}{s}$$

$$= \frac{1 - e^{-as}}{as}$$

Note that the area of the pulse is $a \times 1/a = 1$. Now consider the area of the pulse to remain fixed, but the duration will tend to zero. As the duration tends to zero, the quantity e^{-as} may be expressed as the first two terms of its Taylor series:

$$\lim_{a \to 0} e^{-as} \approx 1 - as$$

Thus in the limit $a \to 0$, the unit pulse will become a unit impulse with Laplace transform:

$$\mathscr{L}(u_0(t)) = 1 \tag{5.12}$$

It should be noted that a unit impulse may produce an unrealistically large system response. As with any other input, it can be scaled suitably.

Sine: We shall see later that the sine function is a useful function for examining the stability properties of a system. It should be observed that both mathematical and physical sine inputs are useful. The sine function is most easily transformed by writing it in its complex exponential form:

$$\sin(\omega t) = \frac{e^{j\omega t} - e^{-j\omega t}}{2j}$$

Because this expression is in the form of an exponential, it can be transformed using the result obtained in Equation 5.2:

$$\mathscr{L}(\sin(\omega t)) = \frac{1}{2j}\left(\frac{1}{s - j\omega} - \frac{1}{s + j\omega}\right) = \frac{1}{2j}\left(\frac{s + j\omega - s + j\omega}{s^2 + \omega^2}\right)$$

$$= \frac{\omega}{s^2 + \omega^2} \tag{5.13}$$

By similar reasoning, it may be shown:

$$\mathscr{L}(\cos(\omega t)) = \frac{s}{s^2 + \omega^2}$$

First Shifting Theorem: There are occasions when all appearances of s are in the form of $(s + a)$. Although there is a mathematical procedure for obtaining time domain functions from Laplace domain function, the process is seldom used in control theory, instead consider:

$$\mathscr{L}(e^{-at}f(t)) = \int_0^\infty e^{-st}e^{-at}f t dt = \int_0^\infty e^{-(s+a)t}f(t)dt$$

We see that the expression is identical to that for a normal Laplace transform except s has been replaced by $(s+a)$ consequently the result may be expressed as:

$$\mathscr{L}(e^{-at}f(t)) = F(s+a) \tag{5.14}$$

An example of this result is:

$$\mathscr{L}^{-1}\left(\frac{(s+a)}{(s+a)^2 + \omega^2}\right) = e^{-at}\cos(\omega t)$$

5.4 INVERSION, THE HEAVISIDE EXPANSION METHOD

The concept of a transfer function, $F(s) = T(s)G(s)$ was introduced and it was indicated that a transfer function in the property of a system and that the response $F(s)$ could be found for different inputs $G(s)$. There are only a limited number of inputs, e.g. step functions, ramp functions, and sinusoidal functions that can be described in manageable $G(s)$ functions. Fortunately a lot of information about system response can be obtained even with this limited range of functions. The concept of a linear system was introduced initially. If a system is linear and the input functions are from the limited set just mentioned, then the form of the output function is:

$$F(s) = T(s)G(s) = \frac{N(s)}{D(s)}$$

where $N(s)$ is the numerator polynomial and $D(s)$ is the denominator polynomial. For *causal* systems:

$$\text{Order } N(s) \leq \text{ Order } D(s)$$

For our purposes:

$$\text{Causal} \equiv \text{Physically realizable}$$

All roots different: Where the roots of $D(s)$ are all different, we can write:

$$F(s) = \frac{N(s)}{\displaystyle\prod_{i=1}^{i=n}(s + a_i)}$$

Assume that F(s) can be written as a sum of *partial fractions*:

$$F(s) = \sum_{i=1}^{i=n} \frac{A_i}{(s + a_i)}$$

Equate these forms and multiply both sides by $(s + a_j)$:

$$(s + a_j) \sum_{i=1}^{i=j-1} \frac{A_i}{s + a_i} + \frac{(s + a_j)}{(s + a_j)} A_j + (s + a_j) \sum_{i=j+1}^{i=n} \frac{A_i}{(s + a_i)} =$$

$$(s + a_j) \frac{N(s)}{\displaystyle\prod_{i=1}^{i=n}(s + a_i)}$$

Or in a more compact notation:

$$A_j + (s + a_j) \sum_{\substack{i=1 \\ i \neq j}}^{i=n} \frac{A_i}{(s + a_i)} = \frac{N(s)}{\displaystyle\prod_{\substack{i=1 \\ i \neq j}}^{i=n}(s + a_i)}$$

Now observe that if $s = -a_j$, then:

$$A_j + 0 = \lim_{s \to -a_j} \frac{N(s)}{\displaystyle\prod_{\substack{i=1 \\ i \neq j}}^{i=n}(s + a_i)}$$

This expression for determining A_j is useful for two reasons. First it allows easy calculation of the A_j terms and secondly it shows that the assumption that $N(s)/D(s)$ could be expressed as a sum of first order partial fractions $A_j/(s + a_j)$ was correct. The expression can be expressed in a more easily comprehensible form:

$$A_j = \lim_{s \to -a_j} \frac{N(s)}{D(s)} (s + a_j) \tag{5.15}$$

where the cancellation of the $(s + a_j)$ is performed in $D(s)$ before taking the limit $s \rightarrow -a_j$. The derivation of the expression for A_j should make it obvious why the method only works for a polynomial $D(s)$ in which there are no repeated roots.

The development up to this point should be giving an indication of the utility of the Laplace transform and the Heaviside method of expanding $F(s) = T(s)G(s)$ into partial fractions. Each partial fraction term in the s or Laplace domain:

$$\cdots \frac{A_j}{s + a_j} \cdots$$

can be inverted by inspection into a time domain response:

$$\cdots A_j e^{-a_j t} \cdots$$

It should be noted that the derivation of A_j is equally valid if a_j is real or complex. Because the coefficients of the polynomials $N(s)$ and $D(s)$ are real for physical systems, then complex quantities must appear as conjugate pairs. These will combine to real second order partial fractions that invert to:

$$B_k e^{-a_k t} \sin \omega_k t \quad \text{and} \quad C_k e^{-a_k t} \cos \omega_k t$$

Repeated roots: Although somewhat rare, repeated roots may occur in the denominator polynomial $D(s)$ in $F(s) = T(s)G(s)$. We shall first examine the special case where the root multiplicity has the form A_i/s^n. The inverse of A_i/s^n will be obtained somewhat circuitously. As explained previously, the formal inverse transform from the Laplace to the time domain is seldom used in control theory. With the benefit of hindsight, consider:

$$\frac{d}{dt} \frac{e^{-st} t^n}{n} = \frac{-s e^{-st} t^n}{n} + e^{-st} t^{n-1}$$

Rearrange this to:

$$e^{-st} t^{n-1} = \frac{d}{dt} \frac{e^{-st} t^n}{n} + \frac{s}{n} e^{-st} t^n$$

Now perform a Laplace transform:

$$\begin{aligned}
\mathscr{L}(t^{n-1}) &= \int_0^\infty \frac{d}{dt} \frac{e^{-st} t^n}{n} dt + \frac{s}{n} \int_0^\infty e^{-st} t^n dt \\
&= \left[\frac{e^{-st} t^n}{n} \right]_0^\infty + \frac{s}{n} \mathscr{L}(t^n)
\end{aligned}$$

$$= [0 - 0] + \frac{s}{n}\mathscr{L}(t^n)$$

$$= \frac{s}{n}\mathscr{L}(t^n)$$

This may be rewritten as the recurrence relationship:

$$\mathscr{L}(t^n) = \frac{n}{s}\mathscr{L}(t^{n-1})$$

Sequentially replace the $\mathscr{L}(t^{n-1})$ terms until $\mathscr{L}(t^0)$ is reached and noting (see unit step result) that $\mathscr{L}(t^0) = 1/s$ thus:

$$\mathscr{L}\left(\frac{t^n}{n!}\right) = \frac{1}{s^{n+1}}$$

Change the power of s to n and inverting this yields the desired result:

$$\mathscr{L}^{-1}\left(\frac{A_i}{s^n}\right) = A_i\frac{t^{n-1}}{(n-1)!} \qquad (5.16)$$

Observe that for the more general repeated root $A_i/(s + a_i)^n$, the First Shifting Theorem can be used to write the inverse as:

$$\mathscr{L}^{-1}\left(\frac{A_i}{(s + a_i)^n}\right) = A_i\, e^{-a_i t}\frac{t^{n-1}}{(n-1)!} \qquad (5.17)$$

Another matter that must be examined is the form of partial fractions associated with the repeated roots. First consider a situation where there is only one repeated factor, $(s + a_j)^r$. All other factors are different. We postulate that the polynomial $F(s) = N(s)/D(s)$ can be written:

$$\frac{A_1}{(s + a_r)} + \frac{A_2}{(s + a_r)^2} + \cdots + \frac{A_r}{(s + a_r)^r} + \sum_{j=1}^{j=n-r}\frac{A_j}{(s + a_j)} = \frac{N(s)}{\displaystyle\prod_{i=1}^{i=n}(s + a_i)} \qquad (5.18)$$

Now multiply both sides by $D(s)$. The orders of the polynomials on the left-hand side will be:

$$(n-1),\ (n-2),\ (n-3), \cdots (n-r),\ \cdots (n-1) \cdots$$

There were r values of A_j associated with the $(s + a_r)^r$ repeated factors. The $(n-1)$ order polynomial will have n coefficients including the constant term. There are n A_j and A_i coefficients so there is enough information to evaluate all the unknown coefficients associated with the $1/(s + a_j)^j$ terms. Thus the proposed partial fraction expansion in Equation 5.18 is valid.

We shall now investigate the formal approach to evaluating the A_j terms for the repeated roots. Write the partial fraction expansion as:

$$\frac{A_1}{(s+a_r)} + \frac{A_2}{(s+a_r)^2} + \frac{A_3}{(s+a_r)^3} + \cdots + \frac{A_r}{(s+a_r)^r} + \theta(s) = \frac{\phi(s)}{(s+a_a)^r}$$

Now multiply both sides by $(s+a_r)^r$:

$$(s+a_r)^{r-1}A_1 + (s+a_r)^{r-2}A_2 + \cdots + (s+a_r)A_{r-1} + A_r +$$
$$(s+a_r)^r\theta(s) = \phi(s) \qquad (5.19)$$

Substitute $s = -a_r$ and the result will be:

$$A_r = \phi(a_r)$$

If the whole expression is differentiated with respect to s and again $s = -a_r$ is substituted into the expression:

$$A_{r-1} = \frac{\phi'(a_r)}{1!}$$

This procedure is repeated until:

$$A_1 = \frac{\phi^{(r-1)}(a_r)}{(r-1)!}$$

Observe that the expression $(s+a_r)^r\theta(s)$ does not need to be differentiated because it will always evaluate to zero when $s = -a_r$.

It should be noted that the process looks deceptively simple. In general it is not, because $\phi(s)$ is the quotient of two polynomials and obtaining the derivative is tedious. Fortunately repeated roots are not common and in most instances where there are two or three repeated roots, a coefficient comparison method is much easier to implement than the differentiation approach.

5.4.1   Repeated Roots in Practice

Before working through an example, this material can be put in context by examining a simple fluid power system. Consider an actuator that is unconstrained by any spring. The actuator is supplied by a spool valve. Suppose the input to the spool valve is a step function. After any transients have decayed, observation would suggest that the velocity of the actuator would become constant and the displacement would become a linearly increasing quantity. From the material covered so far, we can see that the denominator of the output $F(s)$ will have two $1/s$ terms. The first is a property of the unconstrained actuator that acts as an integrator and the second comes from the input which is $G(s) = C_1(1/s)$ where C_1 is simply a numerical factor corresponding to the displacement of the spool valve. The $1/s$ factor would lead to a constant output and the $1/s^2$ factor would lead to a C_2t output. Remembering that outputs are additive for a linear system, then the long term output will be the linear continuously increasing displacement.

5.4.2   Worked Example of Inversion

Consider the Laplace function:

$$F(s) = \frac{10s^4 + 83s^3 + 247s^2 + 324s + 168}{s^5 + 11s^4 + 45s^3 + 85s^2 + 74s + 24}$$

A polynomial root finder will show that the the roots of the denominator are:

$$s = -1 \text{ (repeated twice)}, -2, -3, \text{ and } -4$$

Thus the function can be expressed in partial fraction form:

$$\frac{A_1}{(s+1)} + \frac{A_2}{(s+1)^2} + \frac{A_3}{(s+2)} + \frac{A_4}{(s+3)} + \frac{A_5}{(s+4)} =$$
$$\frac{10s^4 + 83s^3 + 247s^2 + 324s + 168}{s^5 + 11s^4 + 45s^3 + 85s^2 + 74s + 24}$$

The coefficients A_5 to A_2 may be evaluated using the standard Heaviside procedure. Start with A_2:

$$
\begin{aligned}
A_2 &= \lim_{s \to -1} \frac{10s^4 + 83s^3 + 247s^2 + 324s + 168}{(s+2)(s+3)(s+4)} \\
&= \frac{10(-1)^4 + 83(-1)^3 + 247(-1)^2 + 324(-1) + 168}{((-1)+2)((-1)+3)((-1)+4)} \\
&= \frac{10 - 83 + 247 - 324 + 168}{1 \times 2 \times 3} = \frac{18}{6} = 3
\end{aligned}
$$

Similarly it may be shown:

$$A_3 = 2, \ A_4 = 3, \text{ and } A_5 = 4$$

Now it is necessary to evaluate A_1. The formal mathematical way would be to use Equation 5.19 and differentiate the function $\phi(s)$. In this instance the numerator of $\phi(s)$ is a 4th order polynomial and the denominator is

3rd order. The amount of algebraic manipulation is very significant and should really be performed with a computer program to avoid arithmetic errors. For this problem it is much simpler to obtain A_1 by comparison of coefficients. The left-hand side must be expanded until it is over a common denominator equal to the right-hand side denominator. The left-hand side numerators must match the right-hand side:

$$
\begin{aligned}
A_1(s^4 + 10s^3 + 35s^2 + 50s + 24) \quad &+ \\
3s^3 + 27s^2 + 78s + 72 \quad &+ \\
2s^4 + 18s^3 + 54s^2 + 62s + 24 \quad &+ \\
3s^4 + 24s^3 + 63s^2 + 66s + 24 \quad &+ \\
4s^4 + 28s^3 + 68s^2 + 68s + 24 \quad &= \\
10s^4 + 83s^3 + 247s^2 + 324s + 168 &
\end{aligned}
$$

The left and right sides must be equal for any value of s thus each of the coefficients of the powers of s must match. For simplicity, use the coefficients of s^4:

$$A_1 + 2 + 3 + 4 = 10$$
$$A_1 = 1$$

Thus we have all the information necessary to invert $F(s)$ into the time domain form:

$$\mathscr{L}^{-1}\left(\frac{1}{(s+1)} + \frac{3}{(s+1)^2} + \frac{2}{(s+2)} + \frac{3}{(s+3)} + \frac{4}{(s+4)}\right) =$$
$$e^{-t} + 3te^{-t} + 2e^{-2t} + 3e^{-3t} + 4e^{-4t}$$

5.5 STABILITY

We now have developed the tools necessary to determine if a system will be stable. The chapter started with a brief comment that a stable system was one in which the output remained bounded. Stability is a property of the system because the form of $F(s) = T(s)G(s)$ shows that the denominators of $T(s)$ and $G(s)$ are simply multiplied together to form $D(s)$. Thus an unbounded output for a bounded input implies a problem with the denominator of $T(s)$, the *system* transfer function. Remembering that this whole exposition is only dealing with linear systems, then the only output components that will be observed are:

$$\cdots \frac{A_j}{(s+a_j)^r} \cdots$$

The superscript r implies that one or more roots may be repeated. It is far more common to have different roots except for some special examples where s^r may appear. It was just stated that either A_j or a_j can be real or complex, but complex numbers must appear in conjugate pairs so the time domain components are limited to:

$$A_i e^{-a_i t} \quad \text{or} \quad A_j t^{r-1} e^{-a_j t} \quad \text{or} \quad B_k e^{-a_k t} \sin \omega_k t \quad \text{or} \quad C_k e^{-a_k t} \cos \omega_k t$$

where *all* the quantities are real. Actually, although these four output types are mathematically correct, they do not seem to tell the whole story. As shown, the reader would tend to believe that output will always decay to zero after sufficient time has elapsed if the a_i, a_j and a_k are positive. It is quite legitimate, however, to have the condition where some exponent terms disappear. Thus other constant outputs are possible, for example, constant velocity, constant acceleration ... outputs may be possible. By similar reasoning, steady state oscillation is possible.

Although a constant ramp or higher order of t will lead to an output that is unbounded in displacement at large time, such situations are *not* treated as *unstable* in control theory. Control theory limits the term *unstable* to results for which a system has one or more *exponentially* increasing outputs. Thus stability requires that a_i, a_j and a_k must be *positive* quantities for stability. This is the same as saying that the *real* parts of of the *roots* of the denominator polynomial of $T(s)$ must *all* be *negative*. An alternative statement, with exactly the same meaning, is that in a stable system, the roots of the denominator polynomial of $T(s)$ must *all* lie in the *left half* of the complex plane.

5.6   BLOCK DIAGRAMS

The block diagram is a link between a physical system and the sets of differential equations that may be written to describe the performance of the system. One might say that the differential equation set is the ultimate distillation of the essence of the system because it is in a form that can be submitted to a general purpose solver to obtain output vs. time. Unfortunately the distillation is usually so severe that it is difficult to obtain much idea of the physical nature of a system from inspection of the set of equations. Conversely, a drawing of a fluid power system or device will often show so much detail that the sequential operations may be difficult to follow. The block diagram is one level of abstraction more than a physical drawing of a system. The functions of parts of the system are more easily seen and the manner in which an input propagates through the system is more easily understood.

We shall show that block diagrams can be drawn first in the time domain for most fluid power systems because the types of individual transfer elements are limited in number. To be specific, many fluid power devices such as actuators and motors can be represented as integrators. These can be depicted as $\int$ in the time domain. Many other devices simply scale a flow quantity and are represented as constant multiplying blocks. The term *flow quantity* needs some amplification. If valve controlled actuator is taken as an example, the input might be an electrical current to a servovalve. After the current passes through a solenoid, the current will have become a valve displacement, the valve displacement will become a pressure, the pressure a force, the force an acceleration of a mass ... thus the flow quantity usually changes as it is traced through a block diagram. A major utility of the block diagram is that it allows the flow of effects to be observed as an input propagates through a system.

The block diagram containing only certain basic blocks may easily be converted into a block diagram in the Laplace domain because the $\int$ blocks become $1/s$ blocks and other blocks remain the same. Once in the Laplace domain, a set of prescribed operations can be performed on the block diagram so the multiplicity of connected blocks can be reduced to one more complicated block describing the manner in which a flow quantity is acted upon as it passes from input to output. This more complicated block is the transfer function of the system.

This a good point to introduce two separate uses of a block diagram. The purpose of this chapter is to introduce linearization, the Laplace transformation, and stability. The block diagram in the Laplace domain that is reduced to one block containing the ratio of two polynomials is no longer used to provide simulation of a system via inversion to the time domain. Readily available computers and algorithms for solving nonlinear ordinary differential equations have made such a use obsolete. On the other hand, the reduced diagram in the Laplace domain is a useful precursor to examining stability using frequency response techniques.

The second use of the block diagram ignores the possibility of conversion to the Laplace domain. Block diagrams can be drawn with many different nonlinear elements such as orifices, deadband in valves, Coulomb friction, etc. There are modeling programs that can analyze such nonlinear block diagrams and formulate the set of nonlinear differential equations for the system. The program will then solve the equation set numerically for system simulation. This topic will not be examined further in this chapter.

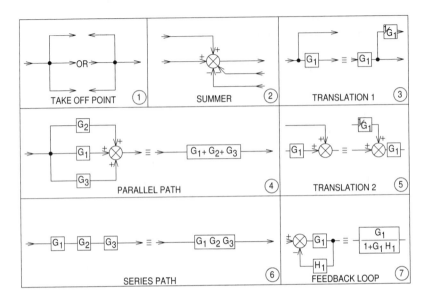

Figure 5.3: Block diagram features.

5.6.1   Consolidation of Block Diagrams

The major features of block diagrams are shown in Figure 5.3. A comment is in order before block diagram consolidation is discussed. We indicated previously that a diagram for a linear system can be written with summers, constant multiplier blocks, and integrators. This is a time domain block diagram and it often very useful when examining interactions between components or the formulation of the system differential equations. It was also stated that such a block diagram in the time domain could easily be converted to a Laplace domain block diagram by replacing all $\int$ blocks by $1/s$ blocks. Observe that consolidation of time domain block diagrams has no meaning. All consolidation to single function blocks is performed in the Laplace domain. The concept of a transfer function only has meaning in the Laplace domain.

With that statement recognized, we can introduce some basic rules that will allow most Laplace domain block diagrams to be consolidated to one, probably algebraically complicated, gain block. This single block is the transfer function between the input and output.

- A flow quantity may be split into multiple paths at a take off point. The amplitude of the flow quantity is the same in each of the multiple paths.

- A summer is used to combine a multiplicity of flow quantities. Note that flow quantities may be added or subtracted at a summer. A multiplicity of entries to one summer may be replaced by a sequence of summers. This is often useful when consolidating feedback loops.

- Translation 1 shows a take off point being moved forward across a gain block G_1. The output from the path after the take off must remain the same so an extra gain block $1/G_1$ must be introduced to achieve this.

- Several parallel paths from a single take off point may be consolidated into one gain block that is the algebraic sum of the individual gain block s functions.

- Translation 2 shows a gain block being forward across a summer. Observe the need to add a $1/G_1$ gain block into the side incoming path to retain the correct output from the summer.

- Gain blocks in series on a common path may be consolidated by multiplication.

- A very common subsection of a block diagram is a feedback loop. In fact, one repeated task in consolidating block diagrams is the inspection of the diagram to see how feedback loops of a more complicated nature can be reduced to the three components shown here. Observe that the side entry to the summer may be additive or subtractive. The entry to the feedback loop will be $U(s)$, here shortened to U. Likewise the output from the loop will be called V. The flow quantity entering the side arm of the summer will be H_1V. The output of the summer will be $U - H_1V$. Consequently the output of the feedback loop may be written as $V = G_1(U + H_1V)$. Performing some simple algebra shows:

$$V = \frac{G_1}{1 + G_1 H_1} U \qquad (5.20)$$

5.6.2 Block Diagram for a Spring-Mass-Damper System

Virtually all moving parts in a fluid power system from the spool in a valve to the gib of a crane that is lifted by a hydraulic actuator are spring-mass-damper systems. The spring may be an oil spring or a separate mechanical

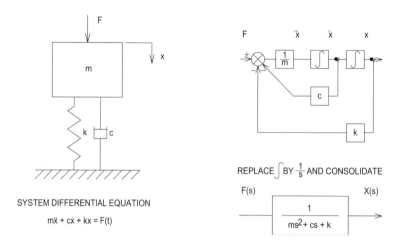

Figure 5.4: Spring-mass-damper, physical form and block diagram.

spring. All terrestrial devices are subject to motion resistance. The resistance may be from shearing of fluid, called viscous damping, or it may be from solid to solid contact, called Coulomb friction. The spring-mass-damper grouping is so fundamental that it will be analyzed in some detail. Normally the assemblages of components in fluid power systems are sufficiently complex that analytical solutions of the dynamic equations are not obtainable. One isolated spring-mass-damper group, however, may be analyzed and an analytical solution found explicitly.

The mass is treated as a point mass, that is, there are no rotational effects. The spring may be mechanical or oil, but its stiffness is considered to be independent of the amount that it is compressed. That is, the increase in force between the ends of the spring is simply proportional to the change in length of the spring at any point in its travel. A real spring has mass, and some of this mass must be associated with the moving mass. In this analysis, the spring is considered massless. For the current analysis, only viscous damping is considered. For the purpose of the analysis, viscous damping means that the force generated by the damper is directly proportional to the velocity difference between the ends of the damper.

Although a spring-mass-damper system is quite simple, its block diagram contains many of the components that will be encountered in the analysis of linearized fluid power systems. Consider the construction of the block diagram in the time domain shown in the upper right corner of

Figure 5.4. The input to the system is a force. This force is reduced by viscous damping $c\dot{x}$ and by spring resistance kx. The resultant force from the input force and the two internally generated forces is established by using a summer. In this instance the two quantities are subtracted from F by attaching minus signs to the arrows entering the summer. An inspection of Newton's Second Law, $F = ma$, shows this can be written as $a = F(1/m)$ thus the functional block between F and $\ddot{x}$ is $1/m$. Conversion of acceleration to velocity is via integration and likewise velocity to displacement. Finally the flow quantities $\dot{x}$ and x are routed from take off points through constant multiplier gain blocks c and k to the summer.

Now convert all the $\int$ blocks to $1/s$ and the time domain block diagram will have been converted to a Laplace domain block diagram. This Laplace domain block diagram may be consolidated into a single gain block. First consider the innermost loop. Applying the rule that blocks in series are multiplied means that the forward path in this innermost loop becomes $1/ms$. This is the gain block G_1 on Figure 5.4. The feedback path on this innermost loop contains the gain block c so $H_1 = c$. Thus applying the procedure for reducing a feedback loop:

$$\frac{G_1}{1 + G_1 H_1} = \frac{1/ms}{1 + c/ms}$$
$$= \frac{1}{ms + c}$$

The original block diagram has now been reduced to a single feedback loop with a forward path gain block of $1/(ms+c)$ and a feedback path of k. Applying the feedback loop consolidation procedure again leads to a single gain block for the system of:

$$T(s) = \frac{1}{ms^2 + cs + k} \tag{5.21}$$

5.7 SPRING-MASS-DAMPER TIME RESPONSE TO UNIT STEP FORCE

Return to the situation shown in Figure 5.4. The applied force can have any variation with respect to time, but only certain functions can be handled analytically. The unit step force will be chosen in this example both because it has a simple transform and because the response shows several interesting features. It was shown in Equation 5.10 that a unit step function could be transformed to $1/s$. It will be recognized that no real force can change instantaneously, but forces that change very rapidly with respect to certain characteristics of the system can effectively be treated as instantaneous.

In order to analyze any system, the initial conditions must be known. For our purposes, this means that the system is at rest and is unchanging with time before the forcing function is applied. If we assume that the displacement x is vertical, then the initial conditions require that the spring is initially compressed somewhat to support the weight mg.

After the force is applied, the motion of the system can be determined by application of Newton's Second Law, thus:

$$F(t) - kx - c\dot{x} = m\ddot{x}$$

where the displacement x is a displacement from the static equilibrium. Because an analytical solution can be obtained, write this equation as:

$$m\ddot{x} + c\dot{x} + kx = F(t) \tag{5.22}$$

With the benefit of hindsight, it turns out that it is worthwhile to express the coefficients of Equation 5.22 in an altered form:

$$\ddot{x} + \frac{c}{m}\dot{x} + \frac{k}{m}x = \frac{F(t)}{m}$$

$$\ddot{x} + 2\zeta\omega_n\dot{x} + \omega_n^2 x = \frac{F(t)}{m} \tag{5.23}$$

Thus ζ and ω_n can be expressed in terms of m, c, and k as:

$$\zeta = \frac{1}{2}c\sqrt{\frac{1}{km}} \tag{5.24}$$

and:

$$\omega_n = \sqrt{\frac{k}{m}} \tag{5.25}$$

Because Equation 5.23 is a second order ordinary differential equation with constant coefficients, it can be solved analytically using the Laplace transform method.

$$\mathcal{L}(\ddot{x} + 2\zeta\omega_n\dot{x} + \omega_n^2 x) = \mathcal{L}\left(\frac{F(t)}{m}\right)$$

$$(s^2 + 2\zeta\omega_n s + \omega_n^2)X(s) = \frac{F(s)}{m}$$

A step function of amplitude F has the transform F/s. Consequently the Laplace domain equation can be rearranged in a form suitable for inversion back into the time domain:

$$X(s) = \frac{F}{m}\left(\frac{1}{s(s^2 + 2\zeta\omega_n s + \omega_n^2)}\right) \tag{5.26}$$

Equation 5.26 can be inverted using the Heaviside inversion method reviewed in Section 5.4.

The denominator of Equation 5.26 has an s term that will invert to a step function and a quadratic. The quadratic is the characteristic polynomial of the system. The roots of this quadratic may easily be found from the standard expression for solving quadratics:

$$\begin{aligned} s &= -\zeta\omega_n \pm \sqrt{\zeta^2\omega_n^2 - \omega_n^2} \\ &= \omega_n(-\zeta \pm \sqrt{\zeta^2 - 1}) \end{aligned} \tag{5.27}$$

Thus there will be two solution regions. If $\zeta < 1$ then the overall solution for s is a pair of complex conjugates. On the other hand if $\zeta \geq 1$, then the overall solution is comprised of two unequal negative real roots.

Solution region where $\zeta < 1$: Equation 5.26 can be rearranged into a form that makes it easier to invert into the time domain:

$$X(s) = \frac{F}{\omega_n^2 m}\left(\frac{1}{s} - \frac{s + \zeta\omega_n}{(s + \zeta\omega_n)^2 + \omega_n^2(1 - \zeta^2)} - \frac{\zeta\omega_n}{(s + \zeta\omega_n)^2 + \omega_n^2(1 - \zeta^2)}\right) \tag{5.28}$$

After some intermediate steps, Equation 5.28 inverts into the form:

$$x(t) = \frac{F}{\omega_n^2 m}\left(1 - \frac{e^{-\zeta\omega_n t}}{\sqrt{1 - \zeta^2}}\sin\left(\omega_n\sqrt{1 - \zeta^2}\,t + \tan^{-1}\frac{\sqrt{1 - \zeta^2}}{\zeta}\right)\right) \tag{5.29}$$

Solution region where $\zeta \geq 1$: Equation 5.26 has two unequal roots in this region and the form before inversion is:

$$\begin{aligned} X(s) = \frac{F}{\omega_n^2 m}\bigg(\frac{1}{s} &- \frac{\sqrt{\zeta^2 - 1} - \zeta}{2\sqrt{\zeta^2 - 1}}\frac{1}{s + \omega_n(\zeta + \sqrt{\zeta^2 - 1})} \\ &- \frac{\sqrt{\zeta^2 - 1} + \zeta}{2\sqrt{\zeta^2 - 1}}\frac{1}{s + \omega_n(\zeta - \sqrt{\zeta^2 - 1})}\bigg) \end{aligned} \tag{5.30}$$

Equation 5.30 inverts into the form:

$$\begin{aligned} x(t) = \frac{F}{\omega_n^2 m}\bigg(1 &- \frac{\sqrt{\zeta^2 - 1} - \zeta}{2\sqrt{\zeta^2 - 1}}e^{-\omega_n(\zeta + \sqrt{\zeta^2 - 1})t} \\ &- \frac{\sqrt{\zeta^2 - 1} + \zeta}{2\sqrt{\zeta^2 - 1}}e^{-\omega_n(\zeta - \sqrt{\zeta^2 - 1})t}\bigg) \end{aligned}$$

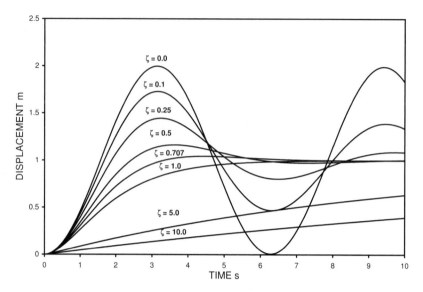

Figure 5.5: Spring-mass-damper, displacement vs. time for step force input.

It may now be more obvious why Equation 5.22 was restated as Equation 5.23. The quantity ζ is a scaled and dimensionless expression of damping that is called the *damping coefficient*. Introducing ζ allows the damping characteristics of any viscous damped system to be described in two numerical ranges, $0 > \zeta < 1$ (underdamped) and $\zeta \geq 1$ (overdamped). At the boundary between an underdamped system and an overdamped system where $\zeta = 1$, the damping is said to be *critical*. The quantity ω_n is the natural frequency of the spring-mass-damper and would be the oscillation frequency were damping to be totally absent. In the underdamped solution range, the solution is oscillatory with frequency $\omega_d = \omega_n \sqrt{1 - \zeta^2}$ where ω_d is the frequency of damped oscillation. Note that this frequency is expressed as radian/second.

Unit values for F, m, and ω_n were used in Equations 5.29 and 5.31 for a range of values of ζ. The results are presented in Figure 5.5. At values of ζ less than 0.5, the oscillatory nature of the response is quite evident. It should be observed that a system with $\zeta = 0$ is probably unattainable for a terrestrial system. Systems in space operating in a vacuum and in zero gravity can show very low values of ζ, but practical fluid power devices will always have $\zeta > 0$.

If a designer has control over the damping coefficient, then it is common to select $0.5 < \zeta < 0.7$. In this range, the overshoot is small, the oscillation damps rapidly, and the rate of rise to the final value is reasonably rapid. The responses shown for $\zeta = 5$ and $\zeta = 10$ are shown to indicate the problems that may result when there is large damping present. Such damping might be present when a viscous damped device is operated in cold weather. Where possible, damping values of $\zeta > 1$ are generally avoided.

An associated benefit of this simulation is that it highlights the effect of the value of the natural frequency, ω_n, on system response. For a given value of ζ, the time for a system to reach some arbitrary fraction of the final response will become less as ω_n increases. Typically values of 90% or 95% of the final response would be used. If the spring rate, k, is largely effected by the oil spring effect (Chapter 2), then maintaining high effective bulk modulus values, β_e should be a goal. Obviously, reducing the mass of the object being moved will also be beneficial. Equations. 5.24 and 5.25 should be reviewed to see the relative effects of m, c, and k on ζ and ω_n.

5.8 TIME CONSTANT

In a fluid power device, elements move under the influence of applied forces. Because elements have mass, the elements take appreciable time to move to their new positions. It is often useful to describe operations in terms of time to achieve some desired motion. It is often not possible to establish exact starting and ending points for motion so it is useful to observe part of the motion and relate this to a descriptive time. One means of doing this is to relate the motion of an element to a *time constant*. This is a concept derived from linear, first order systems, but the concept can often be used approximately for the nonlinear systems found in fluid power systems.

The concept of a time constant will be developed by analyzing the object shown in Figure 5.6. A rigid container of fixed volume has a supply through a small bore tube and the discharge from the volume is through an identical tube. Flow through both tubes is assumed to be laminar. Laminar flow is needed to develop a linear model that can be solved analytically.

Equation 3.8 can be rearranged to describe the flow into the volume and the flow out:

$$Q_i = \frac{(p_s - p)\pi d^4}{128\mu\ell} \quad \text{and} \quad Q_o = \frac{(p - 0)\pi d^4}{128\mu\ell}$$

The rate of change of pressure in the volume may be determined using Equation 4.2:

$$\frac{dp}{dt} = \frac{\beta}{V}(Q_i - Q_o)$$

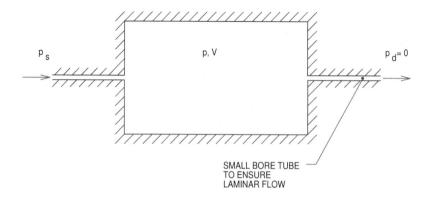

Figure 5.6: Fixed volume with laminar flow connections.

$$= \left(\frac{(p_s - p)\pi d^4}{128\mu\ell} - \frac{p\pi d^4}{128\mu\ell} \right)$$
$$= -2 \left(\frac{\beta}{V} \right) \left(\frac{\pi d^4}{128\mu\ell} \right) (p - \frac{p_s}{2}) \qquad (5.31)$$

The initial conditions are that there is no flow through the chamber and the pressure in the chamber is zero. At time zero, some possibly time varying pressure $p_s(t)$ is applied to the the input. In fact, this pressure will be a step change from 0 to p_{smax}, but treat it as some general input function initially. In this way, an expression of the form $P(s) = T(s)P_s(s)$ can be derived and the transfer function of the system developed. For convenience in equation manipulation, introduce the quantity a where:

$$a = 2 \left(\frac{\beta}{V} \right) \left(\frac{\pi D^4}{128\mu\ell} \right)$$

Thus Equation 5.31 can be written as:

$$\frac{dp}{dt} = -a \left(p - \frac{p_s(t)}{2} \right)$$

This can be reorganized as:

$$2\frac{1}{a}\frac{dp}{dt} + 2p = p_s(t)$$

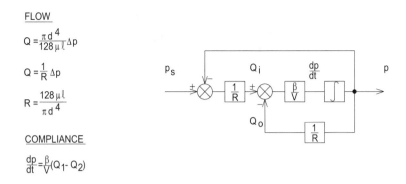

FLOW

$$Q = \frac{\pi d^4}{128\mu l}\Delta p$$

$$Q = \frac{1}{R}\Delta p$$

$$R = \frac{128\mu l}{\pi d^4}$$

COMPLIANCE

$$\frac{dp}{dt} = \frac{\beta}{V}(Q_1 - Q_2)$$

Figure 5.7: Fixed volume with laminar flow connections, time domain block diagram.

Perform a Laplace transform remembering that all conditions are zero before $t = 0$:

$$P(s)\left(\frac{2}{a}s + 2\right) = P_s(s)$$

In transfer function form:

$$P(s) = \frac{a/2}{s+a}P_s(s) \qquad (5.32)$$

This transfer function may also be derived from examination of the block diagram of the device shown in time domain form in Figure 5.7. The equations shown to the left of the block diagram indicate how the flow quantities are transformed as they pass through gain blocks. The gain block between pressure and flow is the reciprocal of resistance. Because the resistance elements are laminar flow tubes, flow is directly proportional to pressure difference. The compliance effect gives a relationship between flow accumulation of a compressible fluid in a volume and the change in pressure. Finally there is a summer to account for difference in pressure driving fluid through the entry resistance and one to establish the accumulation in fluid because of the difference between flow in and flow out of the volume. If the block diagram is converted to the Laplace domain by transforming the $\int$ into $1/s$ and the block diagram reduction rules are applied, then the consolidated gain block matches that in Equation 5.32.

The concept of a time constant is most easily demonstrated if the input function is a step change in pressure. Hence the input function $P_s(s)$ is:

$$P_s(s) = \frac{p_{smax}}{s}$$

Thus the output function that must be inverted is:

$$P(s) = p_{smax}\frac{a}{2}\frac{1}{s(s+a)}$$

In this example, the polynomial ratio is very simple and may be split into partial fractions by inspection by first substituting $s = 0$ and then $s = -a$:

$$P(s) = \frac{p_{smax}}{2}a\left(\frac{1/a}{s} + \frac{-1/a}{s+a}\right)$$

$$= \frac{p_{smax}}{2}\left(\frac{1}{s} - \frac{1}{s+a}\right)$$

Invert this expression back to the time domain by recognizing the standard forms $\mathscr{L}^{-1}(1/s) = 1$ and $\mathscr{L}^{-1}(1/(s+a)) = e^{-at}$:

$$p(t) = \frac{p_{smax}}{2}\left(1 - e^{-at}\right)$$

The time constant is introduced as $\tau = 1/a$ thus:

$$p(t) = \frac{p_{smax}}{2}\left(1 - e^{-t/\tau}\right) \tag{5.33}$$

In this example τ has the value:

$$\tau = 0.5\frac{V}{\beta}\frac{128\mu\ell}{\pi d^4} \tag{5.34}$$

In order to present the results more clearly, the value of the chamber volume, V, was calculated by selecting a unit value for the time constant τ. A check on the evaluation of τ shows that the dimension is time:

$$\tau = 0.5\frac{0.5498 \text{ m}^3}{1.4E{+}9 \text{ Pa}}\frac{128 \times 20.0E{-}3 \text{ Pa}\cdot\text{s} \times 0.1 \text{ m}}{\pi \times 0.002^4 \text{ m}^4} = 1.0 \text{ s}$$

The results of this simulation are shown in Figure 5.8. and are largely what would be expected. Initially there is a rapid rise of pressure in the chamber because the inlet pressure difference greatly exceeds the outlet pressure difference. Ultimately equilibrium is reached, theoretically at infinite time, when the pressure in the chamber is half the supply pressure. It should be obvious, that the equilibrium chamber pressure can be any value between p_s and zero by altering the relative resistances at inlet and outlet.

The major reason for presenting this material is not so much the phenomenon of the rate of buildup of pressure in a chamber so much as the

Table 5.1: Fixed volume with laminar flow connections, example
characteristics

Characteristic	Value	Units
Entry/exit tube length, ℓ	0.1	m
Entry/exit tube diameter, d	0.002	m
Chamber volume, V	0.5498	m^3
Supply side pressure, p_s	2.0E+3	Pa
Drain side pressure, p_d	0	Pa
Effective bulk modulus, β_e	1.4E+9	Pa
Absolute viscosity, μ	20.0E−3	Pa·s

form of the result. In the development of the response of a spring-mass-damper to the step input force shown in Section 5.7, the displacement was shown to be either a product of an exponential function and a sinusoid yielding a decaying oscillation or the sum of two unequal exponentials.

Although fluid power systems seldom have explicit analytical results, the simulations often show strong similarities to results from linear models that do have analytical solutions. Displacements start at initial values and then rise towards final values along curves that resemble exponentials. With a true exponential result as shown in Figure 5.8, the output will be 62.3% of the ultimate output at 1 time constant. Values for 2 and 3 time constants are shown on the figure. Knowing that a device has a short time constant, for example for a directional control valve controlling a massive actuator, may mean the designer can considerably simplify the analysis. The dynamics of valves can be replaced by steady state approximations because the valves will reach their equilibrium positions well before the more massive components.

An earlier comment was made in this section that fluid power components could be characterized by their time constants even if the components could not be modeled analytically. In many instances, experimental or simulation results will yield displacement, velocity, or pressure results that are exponentiallike and approximate values for 62.3%, 86.5%, and 95% response may be measured and approximate time constants associated with specific components. As indicated when discussing the spring-mass-damper simulation results, a designer may need a 95% response in a given time. Such a response will not be possible if the time constant for the device is too large.

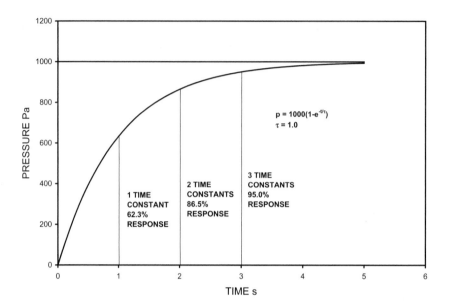

Figure 5.8: Fixed volume with laminar flow connections, pressure vs. time as demonstration of time constant.

PROBLEMS

5.1 Find the time response of the Laplace transform:

$$L(s) = \frac{s + 0.2}{s^2 + 0.4s + 25.04}$$

over a duration of 5× the periodic time. Plot this with any suitable program. Determine the extreme positive amplitude of the response when a time equal to the periodic time has elapsed. Determine a new value of the time constant if the amplitude after one period is to be 0.368. Comment on the relation between time constant and periodic time if the response of an oscillatory system is to be damped rapidly.

5.2 Earlier in the chapter, the comment:

> It will be recognized that no real force can change instanta-
> neously, but forces that change very rapidly with respect to
> certain characteristics of the system can effectively be treated
> as instantaneous.

was made. This problem will allow you to express this statement in
quantitative terms.

Consider the first order differential equation:

$$\frac{dx}{dt} + px = f(t)$$

Obtain the Laplace transform of this equation and express this in the
form:

$$X(s) = T(s)F(s)$$

Now consider the response of this system to a true unit impulse and
to a pulse of amplitude $1/a$ and duration a. Consider the response
to the pulse function at $t = a^-$, that is a minute amount of time
before $t = a$. Suppose this response is k where $0 < k < 1$. Use
a Taylor series expansion of e^{-pt} up to the t^2 term to show that
$a = 2(1 - k)(1/p) = 2(1 - k)\tau$.

If $p = 5$ and $k = 0.99$, generate superposed plots of the impulse and
pulse responses and a plot of the difference between the two responses
divided by the true impulse response. That is you are to plot the
relative error in the response for the second plot. The total elapsed
time for the response curves should be 5 time constants. Repeat the
plots for $k = 0.975$ and $k = 0.95$. Comment on the results with regard
to potential equipment testing.

5.3 Consider a cylindrical piston that is massless. The piston has a small
radial clearance in the bore of a cylinder with the far end of the
cylinder closed. The cylinder is full of oil. To simplify this problem,
you may ignore the effects of gravity and any leakage past the piston.
The piston is given an impulse, e.g. by dropping a steel ball on it and
allowing a rebound. The oil trapped between the perimeter of the
piston and the cylinder wall provides viscous damping. The radial
clearance is very small so laminar flow conditions may be assumed.
The characteristics of the system are given in the table.

Characteristics of a simple system excited by an impulse

Characteristic	Size	Units
Oil bulk modulus, β	0.7E+9	Pa
Oil density, ρ	840	kg/m^3
Oil viscosity, μ	10.0E$-$3	Pa $\cdot$ s
Piston diameter, d	0.1	m
Piston length, ℓ_{pist}	0.2	m
Oil column length, ℓ_{oil}	1.0	m
Radial clearance, δ	25.0E$-$6	m
Mass attached to piston, m	50.0	kg
Impulse forcing function	0.01	N $\cdot$ s

Formulate the equation of motion for the piston and attached mass. Perform a Laplace transform on the equation of motion. If the applied force is in the form of an impulse, invert the Laplace transform to find the response of the system in the time domain.

Use a mathematical program or spreadsheet to generate a table of piston displacement vs. time for a duration equal to 10× the periodic time. Make a plot of the table. Also use the program to calculate the time constant of the system, the natural frequency, the damped frequency, and the damping coefficient.

Repeat the calculations for the extra five sets of values shown in the table.

Mass kg	Viscosity Pa $\cdot$ s		
50	10.0E$-$3	100E$-$3	1.0
5	10.0E$-$3	100E$-$3	1.0

Comment on the change in damping coefficient with the change in values of mass and viscosity.

REFERENCES

1. Wylie, C. R. and Barrett, L. C., 1995, *Advanced Engineering Mathematics*, 6th ed., McGraw-Hill, New York, NY.

6

FREQUENCY RESPONSE AND FEEDBACK

6.1 INTRODUCTION

The concept of frequency response will be covered and it will be shown how this tool can prove a very practical approach to blending control theory and experimental results. The chapter will end by showing how a previous example, the servo controlled actuator having been translated from a physical system to differential equations, can be taken further to a block diagram where the components have physical meaning and functional connectivity. The frequency response approach will be used to obtain an initial estimate of the feedback that can be employed.

The topic of stability was introduced in Section 5.5. It was indicated that the denominator of the transfer function must only have roots in the left hand half of the complex plane. If a system has controlled feedback, then the magnitude of this feedback will generally affect the location of the roots. There are several strategies that may be used to select feedback gain so that a system remains stable and these may be examined in automatic controls texts [1]. Only one method will be reviewed in this chapter, the method of frequency response. We are not claiming that this method is the best method of determining feedback gain, but it does have several attractive features.

- The method can be applied to a mathematical model.

- The method allows gains to be determined by measuring certain dimensions on response plots.

- The method can be applied experimentally to a physical device that can be excited with a varying sinusoidal input.

6.1.1 Heuristic Description

Before starting the mathematical derivation of the method, we shall give a heuristic description. Consider a car traveling at constant speed over a road with a sinusoidal undulation. Personal experience will inform you that if the peaks of the undulations are far apart, then the body of the car will follow the shape of the road very closely. As the wavelength of the undulations is decreased, i.e., the frequency increases, the wheels will follow the road surface, but the body displacement will gradually lessen. Although not obvious to the driver, the car body is showing another response. At low frequencies, the car body displacement is essentially in phase with the road undulations. As the frequency increases, the response of the body lags behind the road displacement.

If a system has deliberate feedback, it will be negative. That is, a positive deviation of output will *reduce* an input valve opening, for example. We would expect this to be desirable otherwise a deviation at the output would cause further valve opening and the system would ultimately have unbounded output. Now consider the negative feedback as the output phase angle lags the input. At some stage, the phase shift might be such that the designed negative feedback could become positive feedback. We shall now formalize these ideas and show that frequency response diagrams can be used to determine how much feedback is acceptable before a system becomes unstable.

6.2 MATHEMATICS OF FREQUENCY RESPONSE

With the previously stated assumptions of linearity, the system transfer function can be expressed as:

$$T(s) = \frac{N(s)}{D(s)}$$

where $N(s)$ and $D(s)$ are polynomials with real, constant coefficients. The system is excited with a sinusoidal input:

$$V(s) = \frac{\omega}{s^2 + \omega^2}$$

Thus the output will be:

$$Y(s) = \frac{\omega N(s)}{(s - j\omega)(s + j\omega)D(s)}$$

It will be assumed that neither $(s^2 + \omega^2)$ nor a similar term $(s^2 + \omega_i^2)$ is a factor of $D(s)$. For fluid power systems, there is always damping, so this assumption is not restrictive. Now express the output in partial fractions:

$$Y(s) = \frac{C_1}{s - j\omega} + \frac{C_2}{s + j\omega} + \left(\frac{A_1}{s + a_1} + \cdots \right)$$

Using the Heaviside expansion method discussed in Chapter 5:

$$\begin{aligned}
C_1 &= \lim_{s \to j\omega} [V(s)T(s)(s - j\omega)] = \frac{\omega T(j\omega)}{j\omega + j\omega} = \frac{T(j\omega)}{2j} \\
C_2 &= \lim_{s \to -j\omega} [V(s)T(s)(s + j\omega)] = \frac{\omega T(-j\omega)}{-j\omega - j\omega} = \frac{T(-j\omega)}{-2j}
\end{aligned}$$

With the assumptions that there are no undamped terms and that the system is stable, it may be stated that all the factors of the form:

$$\left(\frac{A_1}{s + a_1} + \cdots \right)$$

will decay to zero after sufficient time has elapsed. Hence the steady state response to a continuous sinusoidal input is given by:

$$\begin{aligned}
y(t)_{t \to \infty} &= \mathcal{L}^{-1} \left(\frac{T(j\omega)}{2j(s - j\omega)} + \frac{T(-j\omega)}{-2j(s + j\omega)} \right) \\
&= \frac{1}{2j} \mathcal{L}^{-1} \left(\frac{sT(j\omega) + j\omega T(j\omega) - sT(-j\omega) + j\omega T(-j\omega)}{s^2 + \omega^2} \right) \\
&= \frac{1}{2j} \mathcal{L}^{-1} \left(j(T(j\omega) + T(-j\omega)) \frac{\omega}{s^2 + \omega^2} + \right. \\
&\qquad\qquad\qquad\qquad \left. (T(j\omega) - T(-j\omega)) \frac{s}{s + \omega^2} \right) \\
&= \frac{T(j\omega) + T(-j\omega)}{2} \sin \omega t + \frac{T(j\omega) - T(-j\omega)}{2j} \cos \omega t
\end{aligned}$$

We have returned to the time domain and we know that the time domain response is entirely real. Thus $T(j\omega)$ and $T(-j\omega)$ must have the form of complex conjugates:

$$T(j\omega) = \text{Re} + \text{Im}\, j \quad \text{and} \quad T(-j\omega) = \text{Re} - \text{Im}\, j$$

where Re and Im are entirely real quantities. Hence we may write:

$$y(t) = \text{Re} \sin \omega t + \text{Im} \cos \omega t$$

and this may be reduced to the form $C \sin(\omega t + \phi)$:

$$
y(t) = \sqrt{\text{Re}^2 + \text{Im}^2} \left(\frac{\text{Re}}{\sqrt{\text{Re}^2 + \text{Im}^2}} \sin \omega t + \right.
$$

$$
\left. \frac{\text{Im}}{\sqrt{\text{Re}^2 + \text{Im}^2}} \cos \omega t \right)
$$

$$
= \sqrt{\text{Re}^2 + \text{Im}^2} \sin \left(\omega t + \tan^{-1} \text{Im}/\text{Re} \right)
$$

Introduce some standard notation for complex numbers, absolute value $|\ |$ and argument $\angle$:

$$
|T(j\omega)| = \sqrt{\text{Re}^2 + \text{Im}^2} \qquad \angle T(j\omega) = \tan^{-1}(\text{Im}/\text{Re})
$$

$$
y(t)_{t \to \infty} = |T(j\omega)| \sin(\omega t + \angle T(j\omega))
$$

It should be noted that the system was excited with a sinusoid of unit amplitude. If a sinusoid of the form $B \sin \omega t$ had been used in this current mathematical development, then the B factor would have appeared multiplying the output amplitude result. Thus the quantity $|T(j\omega)|$ is called the *amplitude ratio*.

We have now shown that the steady state response of a linear system to a sinusoidal input is a sinusoid of the same frequency. In general, however, the amplitude and phase angle of the output will differ from the input. There is some standard practice in the controls area that should be noted. In conventional mathematical notation, an angle described by radius vector centered at the origin of a set of Cartesian axes is measured in an anticlockwise direction from the positive X axis. Consider an angle of $330°$. This angle could also be denoted as $-30°$. In controls nomenclature, a system displaying such a phase angle would be described as having a phase lag of $30°$.

6.3   FREQUENCY RESPONSE DIAGRAMS

Before discussing the mechanics of constructing frequency response diagrams, we should indicate that these diagrams have the alternative name of Bode plots.

Digression on logarithmic scales: It will be seen that amplitude ratio and frequency are usually presented on logarithmically scaled axes. Although some controls engineers use amplitude ratio axes directly scaled in base 10 logarithms, it is very common to see axes scales in decibels. If the reader has encountered this unit at all, it has probably been in connection with sound. The decibel's application to controls and fluid power deserves some explanation.

The decibel is an artifact from early work on electronic amplifiers for telephone systems. It turns out that the human ear is sensitive to ratios of sound power change so the telephone engineers wanted a unit that described changes in terms of ratios. Consider:

$$P_1 = 10 \text{ W}, \quad P_2 = 100 \text{ W}, \quad P_3 = 1000 \text{ W}$$

The power ratios will be:

$$\frac{P_2}{P_1} = 10 \text{ and } \frac{P_3}{P_2} = 10$$

Now examine the differences in logarithms of these powers:

$$\log_{10} 100 - \log_{10} 10 = 2 - 1 = 1 \text{ and } \log_{10} 1000 - \log_{10} 100 = 3 - 2 = 1$$

So using the logarithmic scaling on the powers would mean that equal changes in logarithms of power would be perceived as equal changes in intensity. Two further factors were involved before the logarithmic scale became the decibel scale. First, the minimum change that can be detected by the average ear is:

$$\log_{10} P_2 - \log_{10} P_1 = 0.1 \text{ or } \log_{10} \frac{P_2}{P_1} = 0.1$$

Second, it was found that for a given loudspeaker system, the change in sound power was a function of current squared, i^2. Combining these two factors meant that the telephone engineers decided that they wanted a minimum unit that could be applied to the easily measured current. This unit was defined as:

$$20 \log_{10} i \text{ measured as decibels}$$

The *bel* part of the unit was a tribute to Alexander Graham Bell, the inventor of the telephone.

The ability to give the units on a logarithmic scale a name was attractive to many controls engineers. It is now common to see amplitude ratios presented in decibels although the quantities being described may be anything from displacements to temperatures.

Frequency response curves for $T(s) = 1/(s + a)$***:*** Make the substitution $s = j\omega$ and evaluate $|T(j\omega)|$ and $\angle T(j\omega)$:

$$T(j\omega) = \frac{1}{j\omega + a} = \frac{-j\omega + a}{\omega^2 + a^2}$$

$$= \frac{1}{\sqrt{\omega^2 + a^2}} \left(\frac{a}{\sqrt{\omega^2 + a^2}} - \frac{\omega}{\omega^1 + a^2} j \right)$$

Observe that:

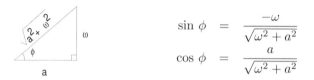

$$\sin \phi = \frac{-\omega}{\sqrt{\omega^2 + a^2}}$$

$$\cos \phi = \frac{a}{\sqrt{\omega^2 + a^2}}$$

thus:

$$|T(j\omega)| = \frac{1}{\sqrt{\omega^2 + a^2}}$$

$$\angle T(j\omega) = \tan^{-1} \left(\frac{-\omega}{a} \right)$$

Now examine the value of $|T(j\omega)|$ for extreme values of ω:

$$|T(j\omega)| = \frac{1}{\sqrt{\omega^2 + a^2}}$$

$$|T(j\omega)| \lim_{\omega \to 0} = \frac{1}{a}$$

$$|T(j\omega)| \lim_{\omega \to \infty} = \frac{1}{\omega}$$

It turns out that the graphs of amplitude ratio vs. frequency have a simpler form when plotted with logarithmic scales. Convert the amplitude ratios at extreme values of frequency to decibels:

$$20 \log_{10} \frac{1}{a} = -20 \log_{10} a$$

and

$$20 \log_{10} \frac{1}{\omega} = -20 \log_{10} \omega$$

At low frequencies the plot is a horizontal straight line passing through $-20 \log_{10} a$ dB and at high frequency the response is a straight line with negative slope descending at 20 dB/decade. The two lines intersect at $\omega = a$. Determine the system amplitude ratio at the asymptote intersection:

$$|T(j\omega)| = \frac{1}{\sqrt{2}\, a}$$

$$20 \log_{10} \frac{1}{\sqrt{2}\, a} = -20 \log_{10} a - 20 \log_{10} \sqrt{2}$$

$$= -20 \log_{10} a - 3.01$$

This result is often expressed in words that the first order response is 3.01 dB down from the low frequency response at the corner frequency. This is the frequency at the asymptote intersection. Incidentally, the frequency axis although logarithmically scaled, will be marked in frequency units.

Now consider the phase angle at low and high frequency:

$$\angle T(j\omega) = \tan^{-1}\frac{-\omega}{a}$$

$$\angle T(j\omega)\lim_{\omega\to 0} = \tan^{-1}\frac{-0}{a} = -0°$$

$$\angle T(j\omega)\lim_{\omega\to\infty} = \tan^{-1}\frac{-\infty}{a} = -90°$$

Also consider the value of the phase angle at $\omega = a$:

$$\angle T(j\omega) = \tan^{-1}\frac{-a}{a} = -45°$$

It will be observed that the phase angles are negative, that is, the output lags the input. Also observe that the angles are generally expressed in degrees and the phase angle axis is linearly scaled.

The development in this section has been for the factor $1/(s+a)$. The denominator factor $1/s$ is commonly encountered. This factor is a single straight line that is the high frequency asymptote with slope $-20\log_{10}\omega$ at all frequencies. The phase angle curve is a horizontal straight line passing through $-90°$. The algebra for numerator terms will not be developed. It is easily shown that $(s+a)$ and s terms have frequency response plots that are mirror images in the frequency axis of the companion denominator terms.

Frequency response curves for $T(s) = 1/(s^2 + 2\zeta\omega s + \omega^2)$:
Consider the general second order curve:

$$T(s) = \frac{1}{s^2 + 2\zeta\omega_n s + \omega_n^2} \tag{6.1}$$

Referring back to Equation 5.27, factor the denominator:

$$s = \frac{-2\zeta\omega_n \pm \sqrt{4\zeta^2\omega_n^2 - 4\omega_n^2}}{2}$$

$$= -\zeta\omega_n \pm \sqrt{\zeta^2 - 1}\,\omega_n \tag{5.27}$$

Write this as:

$$\left(s + \zeta\omega_n + \sqrt{\zeta^2 - 1}\,\omega_n\right)\left(s + \zeta\omega_n - \sqrt{\zeta^2 - 1}\,\omega_n\right)$$

Inspection of these factors shows that for stable systems:

$$\omega_n > 0 \text{ and } \zeta \geq 0$$

An initial discussion of the significance of the value of the damping coefficient, ζ, was introduced in Section 5.7, but the discussion will be extended later in this section.

Following the procedure for the first order system, we shall first examine the characteristics of the system at low and high frequency. Perform the substitution of $s = j\omega$ into Equation 6.1:

$$
\begin{aligned}
T(j\omega) &= \frac{1}{-\omega^2 + 2\zeta\omega\omega_n j + \omega_n^2} \\
&= \frac{1}{(\omega_n^2 - \omega^2) + 2\zeta\omega\omega_n j} \\
&= \frac{(\omega_n^2 - \omega^2) - 2\zeta\omega\omega_n j}{(\omega_n^2 - \omega^2)^2 + (2\zeta\omega\omega_n)^2}
\end{aligned}
$$

Put this expression in standard complex number form $r(\sin\phi + j\cos\phi)$:

$$
T(j\omega) = \frac{1}{\sqrt{(\omega_n^2 - \omega^2)^2 + (2\zeta\omega_n\omega)^2}}
$$
$$
\left(\frac{(\omega_n^2 - \omega^2)}{\sqrt{(\omega_n^2 - \omega^2)^2 + (2\zeta\omega_n\omega)^2}} - \frac{2\zeta\omega_n\omega}{\sqrt{(\omega_n^2 - \omega^2)^2 + (2\zeta\omega_n\omega)^2}} j \right)
$$

Thus the amplitude ratio in decibels is:

$$
A_{dB} = -20\log_{10}\sqrt{(\omega_n^2 - \omega^2)^2 + (2\zeta\omega_n\omega)^2}
$$

Convert the amplitude ratios at extreme values of frequency to decibels:

$$
\begin{aligned}
A_{dB}\left(-20\log_{10}\sqrt{(\omega_n^2 - \omega^2)^2 + (2\zeta\omega_n\omega)^2}\right)\lim_{\omega\to 0} &= -20\log_{10}\sqrt{\omega_n^4} \\
&= -40\log_{10}\omega_n
\end{aligned}
$$

and

$$
\begin{aligned}
A_{dB}\left(-20\log_{10}\sqrt{(\omega_n^2 - \omega^2)^2 + (2\zeta\omega_n\omega)^2}\right)\lim_{\omega\to\infty} &= -20\log_{10}\sqrt{\omega^4} \\
&= -40\log_{10}\omega
\end{aligned}
$$

At low frequencies the plot is a horizontal straight line passing through $-40\log_{10}\omega_n$ dB and at high frequency the response is a straight line with negative slope descending at 40 dB/decade.

Unfortunately the amplitude ratio behavior near $\omega = \omega_n$ cannot be described as easily for second order systems as it can for first order systems. Consider the expression for the amplitude ratio of a second order system:

$$|T(j\omega)| = \frac{1}{\sqrt{(\omega_n^2 - \omega^2)^2 + (2\zeta\omega_n\omega)^2}} \tag{6.2}$$

Experience shows that lightly damped second order systems show resonance effects. That is the amplitude increases significantly at a certain frequency. Any stringed musical instrument would be an example of this phenomenon. This effect can be quantified by differentiating Equation 6.2 with respect to ω and finding the turning point where the derivative is zero. Only the result will be presented here. The value of ω at which $d|T(j\omega)|/d\omega = 0$ is:

$$\omega = \sqrt{1 - 2\zeta^2}\,\omega_n \tag{6.3}$$

Only real frequencies are of interest, so Equation 6.3 shows that the amplitude ratio will only increase near $\omega = \omega_n$ if:

$$0 \leq \zeta < \frac{1}{\sqrt{2}}$$

We should like to observe that the resonance effect may have considerable consequences for the designer of fluid power equipment. Small valves that depend upon viscous damping may show quite low values of damping coefficient ζ. Such valves may oscillate with large amplitude relative to the valve openings that they control. Such oscillation may produced unwanted oscillation of the devices that the valves control. Another effect that is often undesirable is large overshoot of a controlled device when such a device is in the form of a spring-mass-damper. It was mentioned in Section 5.7 that lightly damped systems show several oscillations when subject to a step force input function. It may be observed in Figure 5.5 that if $\zeta = 0.707$, then there is no overshoot. This matches the result just found because $\zeta = 1/\sqrt{2} = 0.707$. In fact, the calculation relating to resonance effects is essentially the same as that required to find the ζ value that just restricts overshoot for a spring-mass-damper system excited by a step input force.

As with first order numerator terms, second order numerator terms are mirror images in the frequency axis of the denominator terms. Also the $1/s^2$ and s^2 terms are similar to the first order terms, but the amplitude ratio slope will be $-40\log_{10}\omega$ (or $40\log_{10}\omega$) and the phase angle $-180°$ (or $180°$).

The amplitude ratio and phase angle curves for first and second order systems are shown in Figures 6.1 and 6.2. Observe that for $\zeta \leq 0.707$, there is no increase in amplitude ratio at the resonant frequency. Also

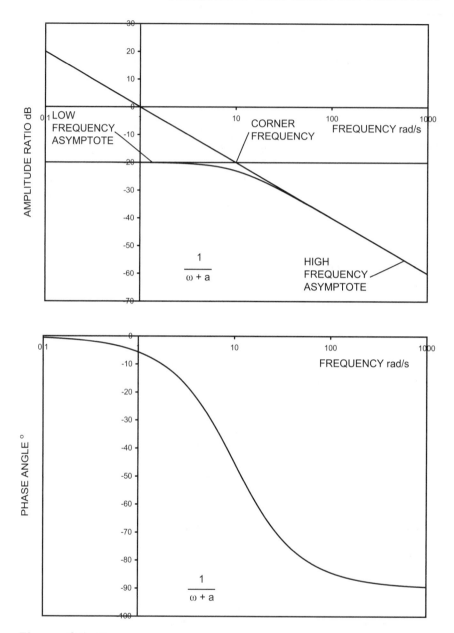

Figure 6.1: Frequency response for first order system.

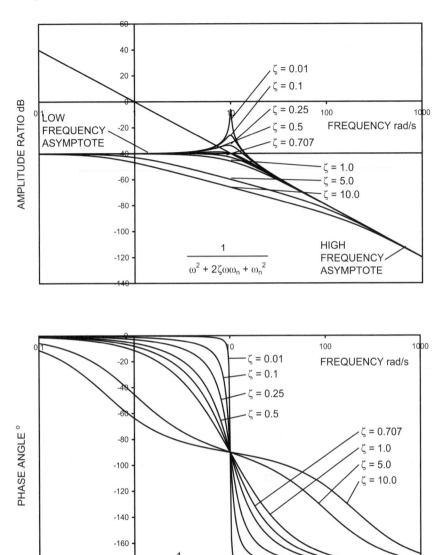

Figure 6.2: Frequency response for second order system.

observe the amplitude ratio when $\zeta = 0.01$ is 34 dB above the low frequency response. This translates to an increase in amplitude ratio of $10^{34/20} = 50.1$. This should be a warning to a fluid power system designer that devices with low damping should be examined carefully to avoid possibly damaging excursions at resonant frequencies.

Comment on higher order polynomial terms: The reader might wonder why so much attention has been given to first and second order polynomials. In fact *only* these terms appear in the evaluation of $T(j\omega)$. Any linear system where $T(s) = N(s)/D(s)$ may be expressed as the ratio of products of only first and second order terms. There will be no higher order polynomials after factoring.

Building curves from components: When $s = j\omega$ is substituted into $T(s)$ each factor in the numerator and the denominator will be a complex number. Note an alternative method of displaying a complex number:

$$x_i + j\, y_i = r_i\, e^{j\phi_i}$$

As indicated in the previous development, the quantity $T(j\omega)$ can be expressed as a product of numerator terms and the inverses of the denominator terms. If logarithmic scaling is used for displaying the amplitude ratio, then the composite amplitude ratio for $T(j\omega)$ will be the product of the r_i terms or the sum of their logarithms. Although the phase angle scale is linear, the same is true for the phase angle because:

$$\cdots\, e^{j\phi_i} e^{j\phi_{i+1}} e^{j\phi_{i+2}}\,\cdots = e^{j(\cdots\,\phi_i + \phi_{i+1} + \phi_{i+2}\,\cdots)}$$

Before the era of easy access to computers, control engineers could construct the amplitude ratio frequency response diagrams with adequate accuracy for many purposes by approximating the individual factor responses with linear asymptotes and hand sketched portions near corner frequencies. Such exercises are no longer necessary, but the concept of summation may be helpful if another factor is added to a system in the form of a controller.

Generating frequency response diagrams: Many mathematical packages can generate frequency response diagrams. If one does not have access to such programs, it is not very difficult to write programs in a general purpose language such as Visual Basic for Applications® for Excel®. The user will have to define a complex variable and provide subroutines for the basic operations such as addition, subtraction, multiplication, and division. It is assumed that the user has the ability to generate

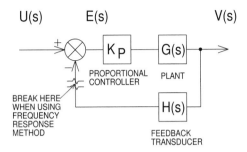

Figure 6.3: Closed loop feedback system showing position of break for conversion to open loop.

frequency response diagrams in an easy fashion when it is remarked later that controller gains are easily determined with little calculation.

6.4   USING FREQUENCY RESPONSE TO FIND CONTROLLER GAIN

A basic feedback control loop is shown in Figure 6.3. Applying the principle given in Equation 5.20 shows that:

$$V(s) = \frac{K_P\, G(s)}{1 + K_P\, G(s)\, H(s)} U(s) \qquad (6.4)$$

If a system contains several components, it is quite likely that there will be some value of K_P that will cause the denominator polynomial $1 + K_P\, G(s)\, H(s)$ to have one or more unstable roots in the right hand half of the complex plane. Although reference to standard controls texts, e.g., [1] will show several methods of finding the value of K_P that just causes instability, the frequency response method has the merit of requiring very little calculation.

Substitute $s = j\omega$ into Equation 6.4 to obtain the frequency response:

$$V(j\omega) = \frac{K_P\, G(j\omega)}{1 + K_P\, G(j\omega)\, H(j\omega)} U(j\omega) \qquad (6.5)$$

An unstable system is one in which the output becomes unbounded even for bounded input. Inspection of Equation 6.5 shows that the output will become unbounded if the denominator becomes zero. In formal terms:

$$1 + K_P\, G(j\omega)\, H(j\omega) = 0$$

or:

$$K_P\, G(j\omega)\, H(j\omega) = -1 \tag{6.6}$$

Now the controller gain K_P and -1 are real numbers, but $G(j\omega)\, H(j\omega)$ is a complex number. Recall that a complex number can be written:

$$x + j\,y = r(\cos\phi + j\,\sin\,\phi)$$

Consequently for a complex number to have a zero imaginary part then:

$$\phi = -(2k+1)\,180°$$

This will cause:

$$\sin(-(2k+1)\,180°) = 0 \ \text{ and } \ \cos(-(2k+1)\,180°) = -1$$

In fact we only need to consider $\phi = -180°$. Thus the stability of a closed loop feedback system can be determined by examining the frequency response of the open loop expression $K_P\, G(s)\, H(s)$. The frequency at which the phase angle curve passes through $-180°$ is noted. The amplitude ratio at this frequency is measured and this must be less than 1 or with the commonly used decibel scale, $A_{dB} \le 0$.

The general procedure is shown on Figure 6.4. The amplitude ratio and phase angle curves are drawn with frequency axes having the same length and the ordinates of the two graph are aligned. A horizontal line is drawn through the $-180°$ mark on the phase angle axis. In general, for systems of order 3 or greater, this line will intersect the phase angle curve somewhere in its right portion.

Draw a vertical line of constant frequency through the value at which the phase angle is $-180°$ and determine where this intersects the open loop frequency response curve that has been drawn with a controller gain $K_P = 1.0$. Inspection of the mathematical development for the frequency response will show that altering the value of K_P has no effect on the phase angle or the *shape* of the amplitude ratio curve, but it will move the amplitude ratio curve up or down. The final position of the amplitude ratio curve should be such that this curve is between 6 and 8 dB below the 0 dB axis when it intersects the constant frequency line just drawn. This value is called the *gain margin*.

After adjusting the amplitude ratio curve to satisfy this condition, determine if the new position of the curve intersects a line through the 0 dB value. The example curve was chosen to do so. Now drop a constant frequency line down from this intersection and determine where it intersects the phase angle curve. For commonly encountered systems, the phase angle on this second constant frequency line will be above the $-180°$ value (i.e. a

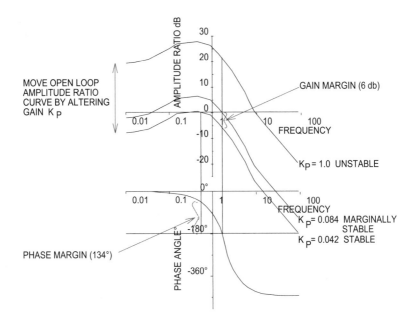

Figure 6.4: Open loop frequency response, determining controller gain.

smaller phase lag). The drop from the phase angle curve to the 180° value is called the *phase margin* of the system.

Standard controls practice uses the criteria:

Gain margin	Phase margin
6 dB	40°
8 dB	30°

As a general rule, do not use less than 6 dB gain margin because there are unavoidable approximations made when linearizing a system, moreover, there will be variations among components. On the other hand, avoid large gain margins because the response will be sluggish.

6.4.1 Example: Hydromechanical Servo Revisited

System analysis: When the hydromechanical servo was simulated in Section 4.4.2, no justification was given for the relative lengths of the levers in the feedback mechanism. We now have the tools to make a formal estimate of these lengths. Examine the schematic given in Figure 6.5 and note that this version of the servo is in the horizontal plane so the effect of an unbalanced gravitational force can be ignored. Apply Equation 4.2 to the oil in the two sides of the actuator:

$$\frac{dp_1}{dt} = (Q_1 - A\dot{x}_a)\frac{\beta_e}{V_1}$$

$$\frac{dp_2}{dt} = (A\dot{x}_a - Q_2)\frac{\beta_e}{V_2}$$

Introduce subscript L for *load* quantities:

$$Q_L = \frac{Q_1 + Q_2}{2} \quad \text{and} \quad p_L = p_1 - p_2$$

The linearized analysis of a servo is performed when the actuator is at the center of its travel (p. 22). This is done because the combined effect of the two oil column springs has its lowest value at this position so the system will be in its least stiff configuration. Such a configuration will be able to sustain less controller gain. Thus introduce:

$$V_{TOT} = V_1 + V_2$$

Combining the two compliance equations and the load definitions yields:

$$\frac{dp_1}{dt} - \frac{dp_2}{dt} = \frac{dp_L}{dt} = (Q_1 + Q_2 - 2A\dot{x}_a)\frac{\beta_e}{V_{TOT}/2}$$

$$= (Q_L - A\dot{x}_a)\frac{4\beta_e}{V_{TOT}} \tag{6.7}$$

The derivation of the method of linearizing the relation between flow, Q_L, valve opening, x_v, and load pressure, p_L is given in Section 7.2.3. The block diagram for the linearized servo with feedback is shown in Figure 6.6.

As might be expected, the diagram looks somewhat like the combination of a spring-mass-damper and a fluid volume. In fact there is no external spring in this assembly and the flow out of the fluid volume is a function of actuator piston velocity. The input to the system is the displacement of the operating lever, x_{op}. This displacement is reduced by feeding back the actuator displacement negatively using the feedback linkage. It was shown in Section 4.4.2, that the valve motion was given by Equation 4.9:

$$x_v = \frac{\ell_2}{\ell_1 + \ell_2}(x_{op} - x_a) \tag{4.9}$$

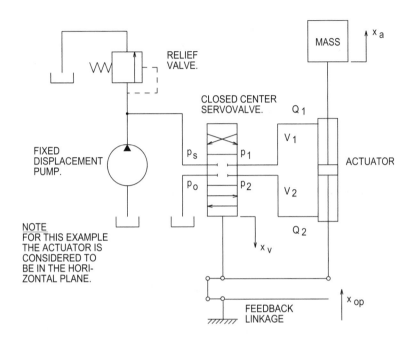

Figure 6.5: Physical diagram of hydromechanical servo with feedback.

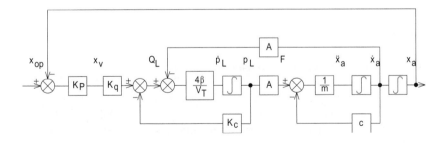

Figure 6.6: Linearized hydromechanical servo, block diagram.

In this development, we have replaced $\ell_2/(\ell_1 + \ell_2)$ by K_P. In a controls text, K_P, would be called *proportional gain* because it alters the corrective input to the system in direct proportion to the error between the input and output. The valve displacement, x_v, is passed through the K_q block and the output will be a flow rate. This flow rate is corrected according to the development in Section 7.2.3 by subtracting a quantity $K_c p_L$. The relationship between flow and the rate of change of pressure in a volume of fluid is given by Equation 6.7. Thus the first part of the block diagram from x_{op}, through the first summer to the output of the first integrator block is a representation of this equation and the linearization of the flow through a valve.

The next part of the block diagram is related to the acceleration of the actuator and its mass according to Newton's Second Law. The load pressure passes through an area multiplying block to yield a force. This force is reduced by the subtraction of the viscous damping force and then acts on the mass. The first integrator in the actuator section of the block diagram converts acceleration to velocity and the second converts velocity to displacement. Both velocity and displacement are needed earlier in the diagram for input into summers, hence the take off points for these quantities.

The last part of the analysis is the formulation of the Newton's Second Law equation for the actuator:

$$\frac{dx_a}{dt} = \dot{x}_a \tag{6.8}$$

$$\frac{d\dot{x}_a}{dt} = \frac{1}{m}(p_L - c\dot{x}_a) \tag{6.9}$$

Although the development in this section has been in the sequence physical layout, block diagram, and then formulation of system equations, it is often easier to conduct the analysis by switching to and fro between the block diagram and the equation formulation paths.

Block diagram consolidation: In order to use the frequency response approach to determine the value of K_P that can be used for stable operation, the integrators in the block diagram must be converted to $1/s$ blocks and the diagram consolidated to the ratio of two polynomials in the Laplace domain. The block diagram has four feedback paths:

$$\dot{x}_a \text{ through } c$$
$$\dot{x}_a \text{ through } A$$
$$p_L \text{ through } K_c$$
$$(x_a \text{ through } 1)$$

None of the first three of these feedback paths have both ends in common. The fourth feedback loop shown in parentheses is from output to input and may be ignored for frequency response analysis because the s function needed is the open loop function with the feedback path broken immediately before the error comparator. Thus three sets of feedback loop consolidation must be performed.

Note that intermediate steps may be made to look simpler algebraically by renaming intermediate consolidated functions. For example, the consolidated function for the next operation will be called $G_1(s)$. The feedback loop for the actuator has a forward gain block of $G(s) = 1/ms$ and a feedback gain block of $H(s) = c$. This loop consolidates to:

$$G_1(s) = \frac{1}{ms + c}$$

Now move the summer at the end of the $K_c p_l$ feedback so that it is ahead of the summer from the $A\dot{x}_a$ path. The consolidation of $G(s) = (4\beta_e/V_{TOT}s)$ and $H(s) = K_c$ is:

$$G_2(s) = \frac{1}{(V_{TOT}/4\beta_e)s + K_c}$$

The third and last feedback loop has a forward path $G(s) = G_2(s)AG_1(s)$ and a feedback path of $H(s) = A$ so this consolidates to:

$$G_3(s) = \frac{1}{(1/G_2(s)AG_1(s)) + A}$$

Thus the complete open loop transfer function in terms of the consolidated $G_i(s)$ terms is:

$$\frac{K_P K_q G_3(s)}{s}$$

Expanding the various $G_i(s)$ terms yields:

$$\frac{K_P K_q A}{\left(\dfrac{m V_{TOT}}{4\beta_e}s^2 + (K_c m + \dfrac{c V_{TOT}}{4\beta_e})s + (cK_c + A^2)\right)s} \tag{6.10}$$

It may be useful to note that s has the dimension T^{-1} so algebra may be checked. All the coefficients multiplied by the appropriate values of T^{-1} must have the same dimensions M, L, T. If similar quantities are being compared at the error comparator, for example here two displacements are being compared, then the dimensions of the final transfer function should be $M^0 L^0 T^0$. This will not always be true, for example the cooling fan for a bus radiator may be driven by a hydraulic motor. The input variable may be temperature and the output motor speed. Comparing these two disparate quantities could be done with transducers that provided electrical outputs. Checking coefficients in the transfer function would be more difficult under such circumstances. Even so, the numerator and denominator dimensions must be individually consistent.

Numerical values: The values for this example are the same as those used in Chapter 4, Section 4.4.1. As mentioned earlier, however, this example assumes that the actuator is operating in the horizontal plane, so weight effects are omitted. These values are presented in Table 6.1.

Table 6.1: Linearized servovalve controlled actuator characteristics

Characteristic	Value	Units
Mass, m	750	kg
Actuator area, A	1.473E−3	m^2
Total oil volume, V_{TOT}	0.7677E−3	m^3
Viscous damping coefficient, c	1.17	N·s/m
Valve area gradient, w	0.01	m
Maximum valve opening, x_{vmax}	0.5E−3	m
Orifice discharge coefficient, C_d	0.62	
Fluid density, ρ	855	kg/m^3
Fluid bulk modulus, β_e	1.2E+9	Pa
System pressure, p_s	20.0E+6	Pa
Drain pressure, p_o	0.0E+6	Pa

The expressions for K_q and K_c, developed in Chapter 7, Section 7.2.3, are repeated here:

$$K_q = C_d w \sqrt{(p_s - p_L)\rho} \qquad (7.4)$$

and:

$$K_c = C_d w \frac{x_v}{2(p_s - p_L)} \sqrt{(p_s - p_L)\rho} = \frac{x_v}{2(p_s - p_L)} K_q \qquad (7.5)$$

The values that the designer can choose for x_v and p_L to calculate K_q and K_c are really unknown until the system design has been completed

and simulated. Obviously not a helpful situation! Thus it will generally
be desirable to estimate the coefficients and perhaps change them if the
simulation gives results that are very different from the initial estimates.
The initial values chosen were $x_v = 0.5$, $x_{vmax} = 0.25E-3$ m and $p_L = 0.25$, $p_s = 5.0E+6$ Pa. Using these values, the two valve coefficients have
initial estimates:

$$K_q = 0.821 \text{ m}^2/\text{s} \text{ and } K_c = 6.84E-12 \text{ m}^3/\text{s} \cdot \text{Pa}$$

The frequency response plots are presented in Figures 6.7 and 6.8. The
initial gain used to plot the open loop amplitude ratio was $K_P = 1.0$. This
gain would lead to an unstable system in closed loop form. Figure 6.7
shows that the controller gain must be reduced by $-22.3 - 6.0 = -28.3$ db
to achieve a stable closed loop system that satisfies both gain and control
margins. This value in decibels translates to $K_P = 10^{-28.3/20} = 0.038$.

Note that Equation 6.10 is fairly generic and may be written:

$$\frac{K_1}{(s^2 + 2\zeta\omega_n s + \omega_n^2)s} \tag{6.11}$$

For this example, $\zeta = 0.16$ and $\omega_n = 134$. The reader now has enough
information that the shape of the frequency response curves for this example
should look familiar. Equation 6.11 has a second order denominator factor
and is underdamped. This factor will have an amplitude ratio curve that
is initially horizontal, shows a peak at $\omega_n = 134$ rad/s, and then declines
along an asymptote with slope -40 dB/dec. The phase angle curve will
start at $\phi = 0°$, it will drop quite sharply through $\omega_n = 134$ rad/s and
then continue at $-180°$. The $1/s$ term has an amplitude ratio that is a
straight line of -20 dB/dec for all frequencies and passes through $\omega = 1.0$
and $A_{dB} = 0$. The phase angle is $-90°$ at all frequencies. Thus this system
will show a phase angle curve that starts at $\phi = -90°$ and descends to
$\phi = -270°$. The total amplitude ratio curve will rise at low frequency so
this system will intersect the $A_{db} = 0$ line somewhere and will allow a phase
margin to be specified.

We should also explain the term *marginal stability*. The definition of
a stable was one with all roots in the left hand half of the complex plane.
The unstable situation occurs with one or more roots in the right hand half.
Thus it is to be expected that there will be a marginally stable situation
where there is at least one pair of roots (i.e. complex conjugates) lying
on the imaginary axis. Thus a system that is marginally stable will show
constant oscillation at constant amplitude. In this example, the gain for
marginal stability is given by $K_P = 10^{-22.3/20} = 0.077$.

Although it is sometimes possible to obtain a numerical solution for
the frequency at which the phase angle is $-180°$, this should not generally

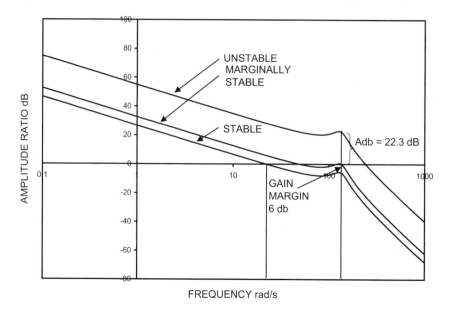

Figure 6.7: Open loop frequency response, hydromechanical servo, amplitude.

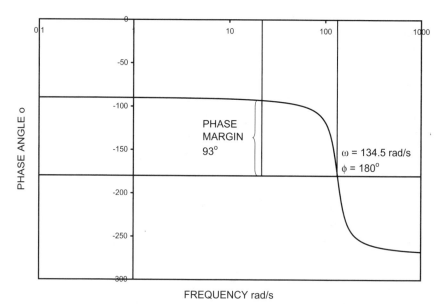

Figure 6.8: Open loop frequency response, hydromechanical servo, phase angle.

be attempted because one of the major advantages of the frequency response method is that it gives usable results with minimal calculation. The derivation of the controller gain that can be used with the closed loop system should be obtained graphically from the plots. The phase margin can even be obtained with quite sufficient accuracy by printing the amplitude ratio curve on thin, translucent paper and sliding the overlay up and down to achieve the required gain margin. The frequency at which the adjusted curve passes through 0 dB can then be marked on the diagrams and the phase margin measured.

Simulation results: It has been stressed on several occasions in this chapter that linear models and the Laplace transform should not be used for simulation. The next results to be presented run counter to that admonition. It may be found in practice, that a marginally stable gain value will not show continuous oscillation when the gain is applied to the original nonlinear system. Figures 6.9 and 6.10 show the *linear* model simulated under marginally stable conditions just to show the reader what might happen. A step input of 0.01 m was applied as $x_{op}(t)$ and the system performance was simulated using Equations 6.7, 6.8, and 6.9.

The reader should now return to Chapter 4, Section 4.4.2 to reexamine the results obtained from the simulation of the hydromechanical servo using a reasonably comprehensive model. The controller gain obtained from the frequency response approach was applied to the servo model from Section 4.4.1 that used viscous damping. The actuator displacement for a step input of 0.1 m from the centered position is shown in Figure 6.11. It will be observed that the response seems fairly adequate, but a small and increasing ripple is observed as time progresses. The selection of values of K_q and K_c was based on guesses for x_v and p_L. Figure 6.12 presents the valve opening for the simulation of the full nonlinear model. The simulated valve opening averaged about 0.002E−3 m. This was about 8× the value used to estimate the linearized valve coefficients. The load pressure is not presented, but it was quite oscillatory. The mean value during the movement to $x_a = 0.1$ m was about 0.7E+6Pa. We should have to admit that the estimate of controller gain obtained from the linearized model and frequency response does not seem very good. Lower gains were investigated and a much smoother response with very little pressure oscillation was obtained if $K_P = 0.02$.

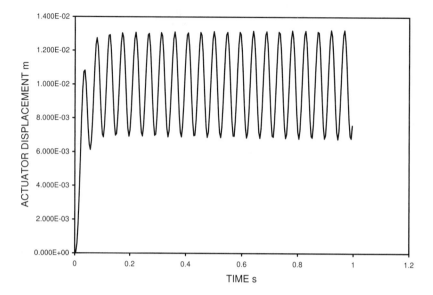

Figure 6.9: Hydromechanical servo linear model, marginal stability, actuator displacement vs. time.

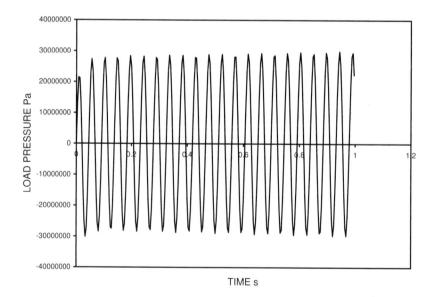

Figure 6.10: Hydromechanical servo linear model, marginal stability, pressure vs. time.

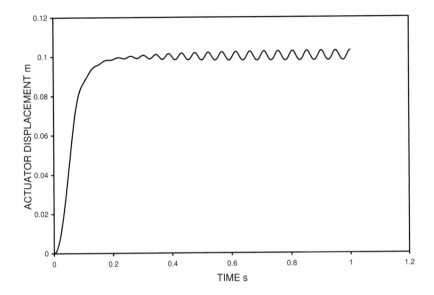

Figure 6.11: Hydromechanical servo nonlinear model, step input, actuator displacement vs. time.

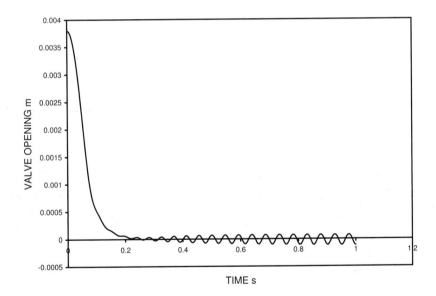

Figure 6.12: Hydromechanical servo nonlinear model, step input, valve displacement vs. time.

6.5   SUMMARY

The discrepancy between the value of gain derived from a linearized model and the frequency response method and that obtained by a cut and try method using a full nonlinear simulation may make the reader feel that the frequency response method is a lot of work for nothing. This conclusion is premature. As was indicated in Section 5.2, nonlinear systems can be linearized near a specific operating point. If the system is operated in a very small range around the operating point, then the linearized model will usually match the more correct nonlinear model quite closely. As the operating range becomes larger, the nonlinearities can be expected to worsen the agreement between the two models. Even if some adjustment is needed before the gain is usable in the nonlinear model, the frequency response method will certainly indicate the order of magnitude of the gain that can be used.

Another feature of the frequency response method is that it can be used with an actual piece of equipment. If a system is too complex to model, it is usually possible to set up variable frequency excitation of the device. The measured open loop frequency response plots may then be used to estimate stable controller gain in exactly the same manner as is done using the mathematical approach.

PROBLEMS

6.1 Use the mathematical program of your choice and generate a frequency response diagram for:

$$F(s) = \frac{2500000}{s^3 + 110s^2 + 251000s + 2500000}$$

Obtain the roots of the polynomial and present this as two simpler polynomials in the denominator. Split the two polynomials into two separate polynomials and examine the material in the chapter and determine what values of numerator constants are required to yield 0 dB amplitude response at low frequencies when each polynomial is plotted separately. Comment on the relationship of the individual frequency response plots to the composite plot.

6.2 The diagram shown in panel (A) of the figure is that of a series resonant circuit and panel (B) shows the block diagram equivalent.

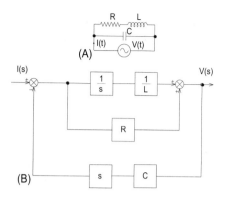

In Chapter 14, it will be shown that there is a close analogy between electrical and fluid circuits when the fluid circuits are excited by sinusoidal variations in flow rate. The potential across the circuit has been labelled $V(t)$ and the flow as $I(t)$ because the reader may be more familiar with the symbols relating to voltage and current.

Reduce the block diagram to a transfer function between the flow rate $I(s)$ and the pressure $V(s)$. Examine the material on spring-mass-damper systems covered in Chapter 5 and develop expressions for natural frequency and damping coefficient for the circuit. Use the values of R, C, and L in the table to determine the natural frequency and damping coefficient for the circuit. Units may appear strange, but they relate to fluid and not the more familiar electrical quantities.

Characteristic	Value	Units
Resistance , R	7.0E+8	$Pa \cdot s/m^3$
Capacitance, C	1.0E−12	m^3/Pa
Inductance, L	4.0E+6	$Pa \cdot s^2/m^3$

Use any available program to plot frequency response diagrams covering the frequency range from $(1/100)\times$ to $100\times$ the natural frequency. Look ahead to Figure 9.12 and observe that axial piston pumps with odd numbers of cylinders have a ripple flow rate of 2×number of pistons×speed. Comment on your results if the RCL system were used on a 5 piston pump running at 477 rpm.

6.3 A common form of control employs elements that work on the error between a desired condition, the input, and the output in such a way that the system receives correction Proportional to error, as the Integral of the error, and as the Derivative of the error. This is a PID controller. Such controller would have the Laplace transform:

$$K_P + \frac{K_I}{s} + K_D s$$

The integral term is used to drive steady state error to zero and the derivative term attempts to control rapid changes in error. Suppose such a controller is used with the system in Problem 6.1. The controller will appear as a term multiplying the $F(s)$ term. Let this term be:

$$K_P \left(1.0 + \frac{0.2}{s} + 0.1s \right)$$

Find the value of K_P that will give a gain margin of 6 dB.

6.4 Rework the example of the hydromechanical servo given in this chapter using the values given in the table. Calculate estimates of K_q and K_c using the same assumptions that were used in the chapter.

Characteristics of a linearized servovalve controlled actuator

Characteristic	Size	Units
Mass, m	75	kg
Actuator area, A	0.247E$-$3	m^2
Total oil volume, V_{TOT}	102.0E$-$3	m^3
Viscous damping coefficient, c	0.38	N $\cdot$ s/m
Valve area gradient, w	0.01	m
Maximum valve opening, x_{vmax}	0.5E$-$3	m
Orifice discharge coefficient, C_d	0.6	
Fluid density, ρ	840	kg/m^3
Fluid bulk modulus, β_e	1.0E+9	Pa
System pressure, p_s	10.0E+6	Pa
Drain pressure, p_o	0.0	Pa

REFERENCES

1. D'Azzo, J. J. and Houpis, C. H., 1988, *Linear Control System Analysis and Design*, 3rd ed., McGraw-Hill Book Company, New York, NY.

7

VALVES AND THEIR USES

7.1 INTRODUCTION

All fluid power circuits incorporate valves. Many authors, e.g., [1], state that there are three categories of valves. These are directional control, pressure control, and flow control. A simple application of a directional control valve would be the valve controlled manually by an operator that determines which end of a cylinder is connected to a pump. The most commonly encountered pressure control valve is the pressure relief valve used to protect components from excess forces caused by overloads or actuators reaching the end of their travel. A flow control valve is used to route oil to a secondary circuit in such a fashion that flow rate remains approximately constant even when pressure is varying.

The principle of operation of most valves is the same. A valve is a variable area orifice where the orifice area may be controlled by conditions in a circuit, for example a pressure relief valve operates without operator intervention. Alternatively the orifice area may be controlled by an operator as in a directional control valve. This categorization is a little simplistic because not all directional control valves are directly linked to an operator. Valves may be moved by electrical actuators and by pressure actuators. Thus many valves are quite complicated in terms of the number of parts in the valve. The valve operation is dynamic and often must be analyzed using methods outlined later in the text (Chapters 11 or 13). Valves may also have feedback loops within them and the stability may have to be examined using control theory.

At the risk of being repetitive, we should briefly revisit the energy equation Equation 3.5. In most valves, fluid at high pressure passes through a pressure drop caused by the orifice. Valves are an indispensable part of

163

most fluid power circuits, but valves usually cause degradation of mechanical energy into heat, so their use leads to a reduction in efficiency of a system. If a system is working in an environment where overall efficiency is very important, for example an aerospace environment, then the designer might want to consider an alternative to a valve. For example, if the desired outcome is a shaft where the speed varies, then there are two options. The flow to the motor can be regulated by a valve fed by a constant pressure system or the flow can be regulated by using a variable displacement pump (i.e., a hydrostatic transmission). The overall efficiency of the hydrostatic transmission will be higher. On the other hand, providing a separate variable displacement pump may add unacceptable cost or weight to the system.

Consider another situation where a pump delivers flow to an actuator and the actuator runs against a stop. The pump must be protected by a relief valve. If a simple relief valve is used, then there will be significant energy degradation because full pump flow passes across a large pressure drop. If an unloading valve is used, then it is possible to arrange that the actuator pressure remains high while the pressure experienced by the pump drops to a low level. In this situation, energy is conserved because the pump flow only passes across a low pressure drop.

The main objective of this chapter is to present several different types of valves and to show what functions these valves can perform, by showing the valves in circuits. The material in the chapter will be largely descriptive. Directional control valves, however, are important components in many circuits. The treatment of such valves has been extended to present quantitative expressions for flow forces on the spool and for linearizing valve performance. The reader should recognize that achieving correct valve operation may require dynamic analysis following the methods presented later in the text (e.g. Chapters 11, 12, and 13).

7.2   DIRECTIONAL CONTROL VALVES

The most common form of directional control valve is the spool valve shown in Figure 7.1. The simplest form of spool is a series of small cylindrical drums on a shaft. Each drum may be called a *land*. The minimum number of lands in a spool valve is two, but four are often used in more expensive valves, such as the proportional type, to achieve more accurate guidance. The valve body has grooves machined in the bore. The edges of the spool lands and the grooves in the bore are machined to a vanishing small radius, so the cylindrical ring orifice that is formed by the displacement of the spool on the bore has sharp edges. It should be observed that fluid passing from

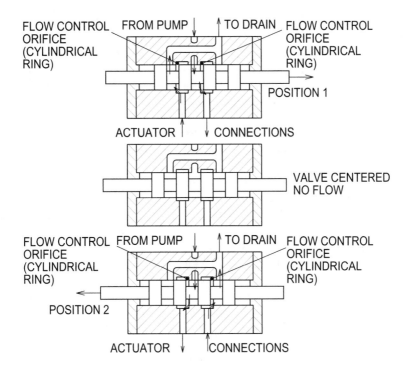

Figure 7.1: Flow through a spool valve.

the pump, to the actuator, and ultimately to the drain, passes through two sequential orifices.

In fluid power terminology, *way* is the name given to an independent passage to or from a valve. Consequently, the valve shown in Figure 7.1 is a *four way*. At first sight, this might seem confusing because there are five passages into or out of the valve. There are, however, two passages to the drain. These two passages may be linked as shown in the figure. An actuator with a spring return only needs one fluid connection. Such a valve could be operated by a three way valve.

The basic spool valve can be produced in several different varieties. In some functions an actuator may only need to be in one of two positions, fully extended and fully retracted. Also, the speed at which the actuator moves may set by the capacity of the circuit supplying the actuator. In this application, the spool valve becomes a switch and only two positions are provided, fully open in one direction and fully open in the other. Although fine machining clearances must still be provided to control leakage, precision control of land and groove width may be relaxed somewhat.

A more demanding application will require a centered position so an actuator may be held in any intermediate position of its travel. The centered position introduces more options. If the lands and grooves are machined so there can be flow from the supply to the drain in the centered position, then the centered position is *tandem centered*. The tandem centered valve has the advantage that fluid can pass from the supply to the drain across a low pressure drop. Consequently, energy loss is limited.

The companion to the tandem center is the *closed center* where flow from supply to drain is blocked in the central position. Both closed and tandem center directional control valves have a small spool movement in the centered position before flow is directed to the actuator. This slack motion is *deadband*.

The most expensive valve available is one in which the groove width and the land width are matched, so leakage flow and deadband at the centered position are minimized. Such valves are called *critical center* valves. Because critical center valves are expensive, they are only used for specialized applications such a servo control of position or velocity. Such an application would be control of flight control surfaces in an airplane.

The options available for directional control valves are initially rather confusing. As with any branch of engineering dealing with assemblies of components into circuits, symbolic representation must provide a clear path of communication between a designer and a builder. We shall not attempt to present all options available, but presenting a few of the basic symbols should help. The basic control valve is represented as two or three adjoining square boxes. Each box represents a possible state of the valve. In the case of the spool valve, one box will show two parallel arrows indicating flow in through one path and return flow via the second path. Another box will show the arrows crossed indicating flow to and from the actuator has reversed. If the valve has a centered position, this may have several options. Only two are shown in Figure 7.2. Flow to and from the supply, the drain, and the two actuator ports is blocked in the closed center position. Flow may take place from the supply to the drain in the tandem center position. Another notation may be observed in the right-hand member of the top panel. Some valves must adjust the position or speed of an actuator in a continuous manner. These proportional control valves are indicated with parallel bars to each side of the possible control positions.

In addition to the center position options, there are the means of moving the spool. The figure shows manual with a lever, a spring, pilot pressure, and a solenoid. It should be realized that the small sample of valve symbols shown in the figure by no means describes all possibilities. For example,

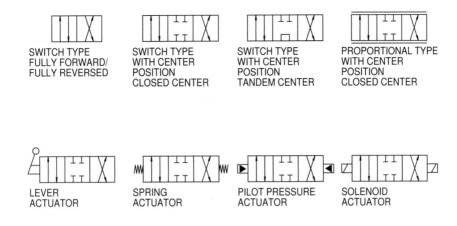

Figure 7.2: Sample directional control valve symbols.

a solenoid operated valve might be returned to an initial position with a spring. Both actuators would be shown on the symbolic representation.

The flow through the control orifices in a spool valve is like that through a nozzle. The change in momentum of fluid flowing through a nozzle requires a force to be applied to the fluid. This force reacts on the spool and is known as the *flow force*. In small valves, the flow force may easily be overcome by the human operator. In larger valves that are operated by solenoids, the flow force may be large enough to require force amplification for the primary actuator. This can be achieved in a pilot operated valve. The solenoid operates on a small spool valve that controls flow to the ends of a large spool valve. That is the large spool is treated as an actuator, but one with a very small displacement. The development of electronic devices has allowed valve manufacturers to build high precision servovalves that are relatively inexpensive. It is now possible to incorporate displacement sensing devices into solenoid valves, so a solenoid does not only supply force, but it can modulate this force according to a position requirement.

7.2.1   Flow Force on a Spool

A control volume for a fluid flowing between two locations is shown in panel (A) of Figure 7.3. For there to be a change in velocity between sections A-A and B-B, a force must be applied to the fluid. Newton's Second Law indicates that force is required to achieve a change in momentum. In general

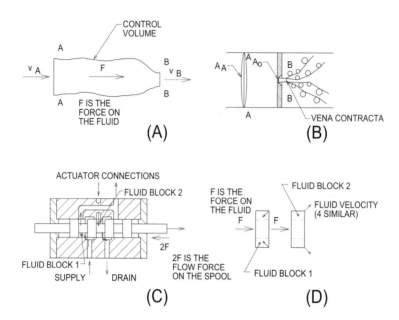

Figure 7.3: Flow forces on a spool valve.

for an undivided stream, we can write:

$$F = \dot{m} v_B - \dot{m} v_A$$

For the specific example of flow forces in a valve, we can consider the magnitude of the velocity downstream of the orifice to be much greater than the upstream velocity. Thus:

$$F \approx \dot{m} v_B$$

Applying conservation of energy from section A-A to B-B and ignoring change in height yields:

$$\frac{v_A^2}{2} + \frac{p_A}{\rho} = \frac{v_B^2}{2} + \frac{p_B}{\rho}$$

Again applying the up and downstream conditions for a small orifice in a large pipe (panel (B)):

$$v_B \approx \frac{\sqrt{2(p_A - p_B)}}{\rho}$$

Accounting for the contraction after the orifice (*vena contracta*) then:

$$
\begin{aligned}
F &= \dot{m}v_B \\
&= (\rho Q)(v_B) \\
&= (\rho C_d A_o v_B)(v_B) \\
&= \rho C_d A_o \frac{2(p_A - p_B)}{\rho} \\
&= 2C_d A_o (p_A - p_B)
\end{aligned}
$$

The angle at which the stream is discharged from the orifice will vary according to the orifice geometry [2]. Based on an angle measured from the spool axis, the angle may be as small as 21° when the spool radial clearance is similar to the spool displacement. The angle may increase to 69° with a large spool displacement. Thus the axial force required to change the momentum of the fluid passing through the valve orifice is:

$$ F = 2C_d A_o (p_A - p_B) \cos\theta $$

The open area of the orifice is proportional to displacement so:

$$ F = 2C_d w x_v (p_A - p_B) \cos\theta \qquad (7.1) $$

Referring to panels (C) and (D) in Figure 7.3, examine the two orifice discharge possibilities. For fluid block 1, the jet is entering the space between two spool lands and is being decelerated. As drawn, the force on the fluid will be from left to right so the reaction force on the spool will be from right to left. For fluid block 2, the fluid in the spool cavity is being injected into the drain port. This fluid is being accelerated and the force on it must be from left to right. As before, the reaction force on the spool is from right to left. Two things should be observed. First, the fluid flow forces are both in the *same* direction. Second, the combined force tends to close the valve. It will be observed from Equation 7.1 that the spool displacement, x_v, appears in the equation, so the flow forces act as a nonlinear spring attempting to center the spool.

One further consequence of flow forces should be mentioned. Because of the jet angle, the flow force may be resolved into two components, the axial component just discussed and a radial component. If the orifice occupies the complete spool perimeter, then the radial component is zero. If only part of the spool perimeter is used, it is important that these parts have symmetry, so the radial force component is zero. Otherwise the nonzero radial component will force the spool against the bore and Coulomb friction forces may cause the spool to stick.

7.2.2   Analysis of Spool Valves

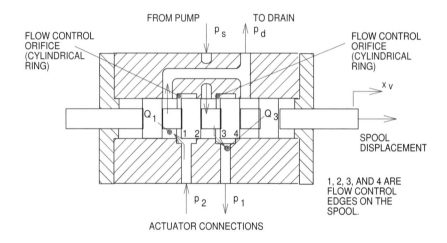

Figure 7.4: Flows and pressures for a servovalve.

Referring to Figure 7.4, we set the supply pressure as p_s and the drain pressure as $p_d = 0$. The pressures on each side of the load are p_1 and p_2 respectively. It is convenient to combine these two pressures in the form of a load pressure, p_L:

$$p_L = p_1 - p_2$$

The major flows through the valve are Q_1 and Q_3. In a real valve, there will be small leakage flows that occur between the spool and the body of the valve. Because this analysis is for an idealized servovalve, all leakage flows in the valve and the actuator will be ignored. Then the flow to the device is:

$$Q_L = Q_1 = Q_3$$

For the type of spool valve of interest, the orifices are rectangular slots when projected on the cylindrical wall of the valve and there is symmetry thus:

$$A_1 = w\,x_v \text{ and } A_3 = w\,x_v$$

where w is called the area gradient. If we assume that the flow through the spool valve is governed by the orifice equation (Equation 3.19), then two equations can be written for the flows:

$$Q_1 = C_d A_1 \sqrt{2(p_s - p_1)/\rho}$$
$$= C_d w x_v \sqrt{2(p_s - p_L)/\rho}$$

and:

$$Q_3 = C_d A_3 \sqrt{2p_2/\rho}$$
$$= C_d w x_v \sqrt{2p_2/\rho}$$

These two flow equations show that:

$$p_s - p_1 = p_2 \text{ or } p_s = p_1 + p_2$$

Now use the load pressure relationship to show:

$$p_1 = \frac{p_s + p_L}{2} \text{ and } p_2 = \frac{p_s - p_L}{2}$$

In other words, as the load is applied, the pressure in one line increases by $p_L/2$ and that in the other decreases by $p_L/2$. Using the expressions among the pressures, to show that the pressure difference across either orifice is:

$$p_s - p_1 = \frac{2p_s}{2} - \frac{p_s - p_L}{2} = \frac{p_s - p_L}{2}$$

Use this form of the pressure difference in the Q_L expression:

$$Q_L = Q_1$$
$$= C_d A_1 \sqrt{(p_s - p_L)/\rho}$$
$$= C_d w x_v \sqrt{(p_s - \frac{x_v}{|x_v|})p_L}$$

The justification for writing $x_v/|x_v|$ may not be immediately apparent. If $x_v > 0$, then $x_v/|x_v| = +1$ and:

$$Q_L = C_d w x_v \sqrt{(p_s - p_L)/\rho} \tag{7.2}$$

alternatively if $x_v < 0$, then $x_v/|x_v| = -1$ and:

$$Q_L = C_d x_v \sqrt{(p_s + p_L)/\rho} = -C_d |x_v| \sqrt{(p_s - |p_L|)/\rho}$$

In the $x_v < 0$ situation, $p_L < 0$, so the term inside the radical will have the right value and the flow will reverse as needed.

Let us manipulate the expression so that it can be displayed in the form of normalized curves:

$$\frac{Q_L}{C_d\, w\, |x_{vmax}|\sqrt{p_s/\rho}} = \left(\frac{x_v}{|x_{vmax}|}\right)\sqrt{\left(1 - \frac{x_v}{|x_v|}\frac{p_L}{p_s}\right)} \qquad (7.3)$$

where x_{vmax} is the maximum opening of the spool. Equation 7.3 has been plotted in Figure 7.5. Servovalves are commonly electrically driven. Manufacturer's literature for such valves will have the shape shown in Figure 7.5. The abscissa will be p_L for some given p_s, the ordinate will be Q_L, and the family of curves will be based on a current ratio, $|i/i_{max}|$, instead of a spool displacement ratio, $|x_v/x_{vmax}|$.

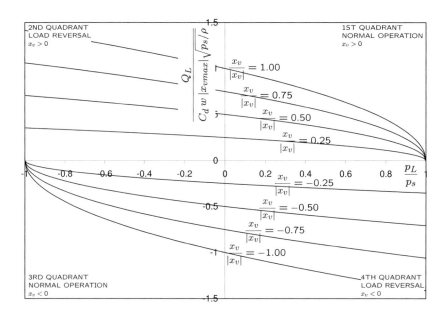

Figure 7.5: Normalized load flow vs. load pressure.

7.2.3   Linearized Valve Coefficients

Basic automatic control theory is based on the manipulation of ordinary differential equations where the coefficients are constant (Chapter 5). The analysis just developed and the analyses presented in Chapter 4 all show that the commonly encountered governing equations and resulting differential equations are nonlinear. It is often possible to linearize fluid power system about some operating point and use automatic control theory on the linearized equations. For example, linearized theory might be used to obtain an initial estimate of the magnitude of feedback that can be used before a system becomes unstable (Chapter 6). This value could then be tried in a more rigorous, nonlinear simulation of a system. The more realistic simulation would allow the magnitude of the feedback to be adjusted.

Fortunately the objective of automatic control is to keep a system close to a specific operating condition. Thus linearization about an operating point is often quite adequate for design. It should be noted that the goal of linearization is not so much the simplification of equations as the obtaining of broader insight into system operation.

The basis of linearization is the expansion of a nonlinear expression using a Taylor series truncated after the first differential term.

$$Q_L(x_v + \Delta x_v, p_L + \Delta p_L) \approx Q_L(x_v, p_L) + \Delta x_v \left. \frac{\partial Q_L}{\partial x_v} \right|_{p_L const} + \Delta p_L \left. \frac{\partial Q_L}{\partial p_L} \right|_{x_v const}$$

The $\partial Q_L / \partial x_v$ coefficient is generally called K_q, the *Flow Gain Coefficient*:

$$K_q = \frac{\partial Q_L}{\partial x_v} = C_d \, w \sqrt{(p_s - p_L)/\rho} \qquad (7.4)$$

The $\partial Q_L / \partial p_L$ coefficient is given a negative sign because the differentiation of Q_L results in a negative result and it is conventional to use K_c, the *Flow Pressure Coefficient*, as a positive quantity. Thus:

$$K_c = -\frac{\partial Q_L}{\partial p_L} = \frac{C_d \, w}{2} \frac{x_v}{p_s - p_L} \sqrt{(p_s - p_L)/\rho} = \frac{x_v}{2(p_s - p_L)} K_q \qquad (7.5)$$

A third coefficient can be derived, the *Pressure Sensitivity Coefficient*, K_p:

$$K_p = \frac{\partial p_L}{\partial x_v}$$

It is often just stated that $K_p = K_q / K_c$, but it will be derived here in an alternative fashion. We believe that this form of derivation shows the relationship to the two independent variables, x_v and Q_L, more clearly. Consider:

$$p_L = f(x_v, Q_L)$$

then:

$$K_p = \left. \frac{\partial p_L}{\partial x_v} \right|_{Q_L const}$$

Write the orifice equation as:

$$\sqrt{(p_s - p_L)} = \frac{Q_L}{C_d \, w \sqrt{\rho}} \frac{1}{x_v}$$

Now differentiate with respect to x_v holding Q_L constant:

$$\frac{1}{2} \frac{1}{\sqrt{(p_s - p_L)}} \left(-\frac{\partial p_L}{\partial x_v}\right) = Q_L \frac{\sqrt{\rho}}{C_d \, w} \left(-\frac{1}{x_v^2}\right)$$

$$\frac{1}{2} \frac{1}{(p_s - p_L)} \frac{\partial p_L}{\partial x_v} = Q_L \frac{1}{C_d \, w \, x_v} \sqrt{\frac{\rho}{(p_s - p_L)}} \frac{1}{x_v}$$

$$\frac{\partial p_L}{\partial x_v} = \frac{2(p_s - p_L)}{x_v} = K_p$$

Thus, as stated earlier:

$$K_p = \frac{K_q}{K_c} = \frac{2(p_s - p_L)}{x_v} \tag{7.6}$$

The feature of this derivation that should be noted is that K_p is a *partial* differential because there were 2 variables on the right-hand side of the equation, x_v and Q_L. In this evaluation, Q_L is assumed constant.

7.2.4   Example: Using the Valve Coefficients

A spool valve is controlling an ideal motor that drives a load having the power characteristic:

$$P = K_{pwr} \, \omega^2$$

We should mention that the relationship between load power and speed will seldom have a simple relationship like that chosen. As will be seen, however, we shall need to know the load power vs. speed relationship because this problem will estimate the change in operating conditions as the speed is changed.

The system has the initial settings given in Table 7.1. Please note two features of this problem. First the ideal motor, the derivation of the linearized model of a servovalve incorporated a no-leakage requirement for the actuator. The second feature involving the power characteristic of the load also turns out to be important. If a load like a hoist were chosen where the pressure requirement is independent of speed, then the approximation using K_q to calculate the new value of Q_L would be identical to the exact calculation.

Table 7.1: Characteristics used to demonstrate the use of valve linearization

Characteristic	Value	Units
Orifice discharge coefficient, C_d	0.62	
Valve area gradient, w	0.01	m
Valve opening, x_v	0.25E−3	m
Supply side pressure, p_s	20.0E+6	Pa
Drain side pressure, p_d	0	Pa
Upstream pressure to motor, p_1	12.0E+6	Pa
Fluid density, ρ	855	kg/m^3
Motor speed, n	1000	rpm

First calculate certain properties associated with the initial conditions. Calculate the load pressure. The valve was indicated to be ideal with no leakage and the motor is also ideal. Thus:

$$p_2 = p_s - p_1 = 20.0\text{E}+6 - 12.0\text{E}+6 = 8.0\text{E}+6 \text{ Pa}$$

So the load pressure can be determined:

$$p_L = p_1 - p_2 = 12.0\text{E}+6 - 8.0\text{E}+6 = 4.0\text{E}+6 \text{ Pa}$$

Calculate the initial load flow:

$$
\begin{aligned}
Q_L &= C_d\, w\, x_v \sqrt{\frac{(p_s - p_L)}{\rho}} \\
&= 0.62 \times 0.01 \times 0.25\text{E}{-}3 \sqrt{\frac{(20.0\text{E}+6 - 4.0\text{E}+6)}{855}} \\
&= 0.212\text{E}{-}3 \text{ m}^3/\text{s}
\end{aligned}
$$

Now use Equations 7.4, 7.5, and 7.6 to evaluate K_q, K_c, and K_p at the initial operating point. The flow gain coefficient, K_q:

$$
\begin{aligned}
K_q &= C_d\, w \sqrt{\frac{(p_s - p_L)}{\rho}} = 0.62 \times 0.01 \sqrt{\frac{(20.0\text{E}+6 - 4.0\text{E}+6)}{855}} \\
&= 0.8481 \text{ m}^2/\text{s}
\end{aligned}
$$

The flow pressure coefficient, K_c:

$$K_c = K_q \frac{x_v}{2(p_s - p_L)} = 0.8481 \frac{0.25\text{E}{-}3}{2(20.0\text{E}{+}6 - 4.0\text{E}{+}6)}$$

$$= 6.626\text{E}{-}12 \text{ m}^3/\text{s} \cdot \text{Pa}$$

The pressure sensitivity coefficient, K_p:

$$K_p = \frac{K_q}{K_c} = \frac{0.8481}{6.626\text{E}{-}12}$$

$$= 0.128\text{E}{+}12 \text{ Pa/m}$$

Calculate the power delivered by the motor:

$$P = p_L \, Q_L = 4.0\text{E}{+}6 \times 0.212\text{E}{-}3 = 848.1 \text{ W}$$

Calculate the motor shaft speed, ω:

$$\omega = \frac{n \, 2\pi}{60} = \frac{1000 \times 2 \times \pi}{60} = 104.7 \text{ rad/s}$$

Calculate the power coefficient K_{pwr}:

$$K_{pwr} = \frac{P}{\omega^2} = \frac{848.1}{104.7^2} = 77.34\text{E}{-}3 \text{ W/rad}^2$$

Calculate the motor displacement:

$$D_m = \frac{Q_L}{\omega} = \frac{0.212\text{E}{-}3}{104.7} = 2.025\text{E}{-}6 \text{ m}^3/\text{rad}$$

Develop a relationship that will allow the calculation of the load pressure, p_L, when the load follows $P = K_{pwr}\omega^2$:

$$P = p_L \, Q_L = K_{pwr} \, \omega^2 = K_{pwr} \left(\frac{Q_L}{D_m} \right)^2$$

Which leads to an expression for Q_L:

$$Q_L = \frac{D_m^2}{K_{pwr}} p_L$$

Noting that Q_L may also be expressed as:

$$Q_L = C_d \, w \, x_v \sqrt{\frac{(p_s - p_L)}{\rho}}$$

Thus Q_L can be eliminated leaving:

$$\frac{D_m^2}{K_{pwr}} p_L = C_d\, w\, x_v \sqrt{\frac{(p_s - p_L)}{\rho}}$$

This expression can be rearranged as a quadratic equation to calculate p_L:

$$p_L^2 + \left(\frac{C_d\, w\, x_v K_{pwr}}{D_m^2}\right)^2 \frac{1}{\rho} p_L - \left(\frac{C_d\, w\, x_v K_{pwr}}{D_m^2}\right)^2 \frac{1}{\rho} p_s = 0$$

Introduce a quantity C_1 where:

$$C_1 = \left(\frac{C_d\, w\, x_v K_{pwr}}{D_m^2}\right)^2 \frac{1}{\rho}$$

For the new operating condition where:

$$x_v = 1.05 x_v = 1.05 \times 0.25\text{E}{-}3 = 0.2625\text{E}{-}3 \text{ m}$$

The value of C_1 is:

$$C_1 = \left(\frac{0.62 \times 0.01 \times 0.2625\text{E}{-}3 \times 77.34\text{E}{-}3}{(2.025\text{E}{-}6)^2}\right)^2 \frac{1}{855} = 1.103\text{E}{+}6$$

Now solve the quadratic in p_L:

$$p_L^2 + C_1\, p_L - C_1\, p_s = 0$$

Only the positive root has meaning in this context, so substituting $C_1 = 1.103\text{E}{+}6$:

$$
\begin{aligned}
p_L &= \frac{-1.103\text{E}{+}6 + \sqrt{(1.103\text{E}{+}6)^2 + 4 \times 1.103\text{E}{+}6 \times 20.0\text{E}{+}6}}{2} \\
&= 4.177\text{E}{+}6 \text{ Pa}
\end{aligned}
$$

Calculate the load flow, Q_L, under the new conditions:

$$
\begin{aligned}
Q_L &= C_d\, w\, x_v \sqrt{\frac{(p_s - p_L)}{\rho}} \\
&= 0.62 \times 0.01 \times 0.2625\text{E}{-}3 \sqrt{\frac{(20.0\text{E}{+}6 - 4.177\text{E}{+}6)}{855}} \\
&= 0.2214\text{E}{-}3 \text{ m}^3/\text{s}
\end{aligned}
$$

Now calculate ω and verify that the predicted p_L is correct by calculating the power under the new conditions in two separate ways:

$$\omega = \frac{Q_L}{D_m} = \frac{0.2214\text{E}{-}3}{2.025\text{E}{-}6} = 109.4 \text{ rad/s}$$

Calculate the power from:

$$P = p_L\,Q_L = 4.177\text{E}{+}6 \times 0.2214\text{E}{-}3 = 924.8 \text{ W}$$

and from:

$$P = K_{pwr}\,\omega^2 = 77.34\text{E}{-}3 \times 109.4^2 = 924.8 \text{ W}$$

Both the power expressions yield the same result. Calculate the increment in valve opening, Δx_v:

$$\Delta x_v = 0.05 \times 0.25\text{E}{-}3 = 0.0125\text{E}{-}3 \text{ m}$$

Now calculate the approximate value of the load flow, Q_L, using K_q from the linearization approach:

$$\begin{aligned} Q_L \approx Q_{Lold} + \Delta x_v\,K_q &= 0.2120\text{E}{-}3 + 0.0125\text{E}{-}3 \times 0.8481 \\ &= 0.2214\text{E}{-}3 \text{ m}^3/\text{s} \end{aligned}$$

and the approximate value of the load pressure, p_L, using K_p:

$$p_L \approx p_{Lold} + \Delta x_v\,K_p = 4.0\text{E}{+}6 + 0.0125 \times 0.128\text{E}{+}12 = 5.6\text{E}{+}6 \text{ Pa}$$

7.2.5   Comments on the Worked Example

The agreement between the exact value of Q_L computed from the power curve and orifice expressions and the estimate obtained by using K_q is quite good. It should be observed that some error could be removed by using the value of Δp_L because the term:

$$\Delta p_L \frac{\partial Q_L}{\partial p_L}$$

is being ignored in the linearized approach to evaluating Q_L. In fact this is not realistic because Δp_L is not known, only estimates are available.

On the other hand, the agreement between the exact value of p_L and its estimate using K_p is rather poor. In this instance, the poor agreement between the two values of p_L can be explained by examining the assumption that Q_L remains constant at the initial and final valve openings. Suppose the relation between Q_L and p_L can be expressed as:

$$Q_L = g(p_L)$$

Rearrange Equation 7.2 as:

$$
\begin{aligned}
\sqrt{(p_s - p_L)} &= Q_L \frac{1}{C_d \, w \sqrt{\rho}} \frac{1}{x_v} \\
&= g(p_L) \frac{1}{C_d \, w \sqrt{\rho}} \frac{1}{x_v}
\end{aligned}
$$

Then expression for K_p can now be derived as an ordinary differential coefficient without making the assumption that Q_L remains constant. The calculus and algebra involved are somewhat tedious and will not be displayed here. The result is:

$$\frac{dp_L}{dx_v} = \frac{2(p_s - p_L)/x_v}{1 + g'(p_l)\,2(p_s - p_L)/Q_L}$$

In the worked example with the power relationship $P = K_{pwr}\omega^2$:

$$Q_L = g(p_L) = \frac{D_m^2}{K_{pwr}} p_L$$

thus in this specific example:

$$g'(p_L) = \frac{d}{dp_L} g(p_L) = \frac{D_m^2}{K_{pwr}}$$

Evaluate the expression dp_l/dx_v for this specific problem:

$$
\begin{aligned}
&\frac{dp_L}{dx_v} \\
&= \frac{2(p_s - p_L)/x_v}{1 + (D_m^2 \, 2(p_s - p_L))/(K_{pwr} \, Q_L)} \\
&= \frac{0.128\text{E}{+}12}{1 + ((2.025\text{E}{-}6)^2 \times 2 \times (20.0\text{E}{+}6 - 4.0\text{E}{+}6))/(77.34\text{E}{-}6 \times 2.214\text{E}{-}4)} \\
&= \frac{0.128\text{E}{+}12}{1 + 4} = 0.0256\text{E}{+}12 \text{ Pa/m}
\end{aligned}
$$

Thus accounting for the variation of Q_L as a function of p_L leads to a prediction of the load pressure under the changed valve opening of:

$$p_L \approx p_{Lold} + \Delta x_v \, dp_L/dx_v = 4.0\text{E}{+}6 + 0.0125 \times 0.0256\text{E}{+}12 = 4.32\text{E}{+}6 \text{ Pa}$$

This is in much better agreement with the exact value calculated using the orifice expression and the load power characteristic.

A similar analysis can be done for the K_q term, that is replacing it by dQ_L/dx_v and allowing for $P = K_{pwr}\,\omega^2$. The value of Q_L predicted by using dQ_L/dx_v is in better agreement with the exact value, but the relative correction is not as great.

In conclusion, linearization may be an essential part of the analysis of some problems. Unquestioning use of the expressions for K_q, K_c, and K_p may lead to satisfactory results. In fact, if no further information is available such as the load power characteristic, that path is the only choice. On the other hand, if more information is available, it should be used.

7.3   SPECIAL DIRECTIONAL CONTROL VALVES, REGENERATION

Some applications require a relatively rapid motion in one direction against a low load followed by a short stroke against a high load. Such an application might be a clamp fixture on a production machine tool. Somewhat large clearance might be required to place the part to be machined in position. The actuator would then be expected to extend rapidly until contact was made with the fixture. At this stage the operator would apply a larger clamping force. A simple method of achieving this sequence is to incorporate regeneration.

Regeneration requires an actuator where the cap side area exceeds the rod side. We shall also see that the system may often require a special form of directional control valve.

During regeneration, flow from the pump that enters the cap side is augmented by adding flow from the rod side. When the large force is required, the flow from the rod side is routed to the drain by action of the directional control valve. Full system pressure is then available over the full cap area. The system is shown in Figure 7.6. During extension, the actuator velocity will be:

$$v_{ext} = \frac{Q + vA_{rod}}{A_{cap}}$$

$$= \frac{Q}{A_{cap} - A_{rod}} \tag{7.7}$$

The force capability will be:

$$F_{ext} = p_s(A_{cap} - A_{rod}) \tag{7.8}$$

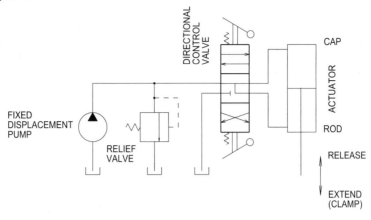

Figure 7.6: Actuator operation incorporating regenerative flow.

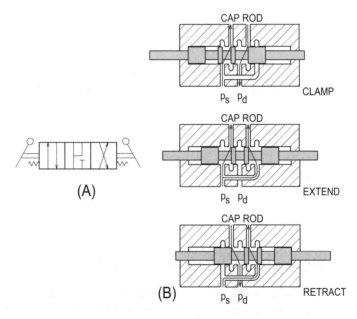

Figure 7.7: Specialized form of directional control valve used when regeneration flow in an actuator is used.

As indicated previously, when the actuator reaches the point in its travel when more force and less velocity is required, the pump flow is directed to the cap area alone and the flow from the rod side flows out of the valve drain port. The velocity is now:

$$v_{clmp} = \frac{Q}{A_{cap}} \tag{7.9}$$

and the force capability is:

$$F_{clmp} = p_s A_{cap} \tag{7.10}$$

During retraction, the pump flow is directed to the rod side and the flow from the cap side flows out of the valve drain port. The retraction velocity will be:

$$v_{ret} = \frac{Q}{A_{rod}} \tag{7.11}$$

and the force capability:

$$F_{ret} = p_s A_{rod} \tag{7.12}$$

The special form of directional control valve is shown in Figure 7.7. The standard symbolic representation is given in panel (A) and the actual configuration of a spool valve in panel (B). If Figure 7.6 is compared to the normal directional control spool valve in Figure 7.1, it may be seen that the spacing of the two innermost lands has been increased so that flow can enter both the cap and rod sides of the actuator when the valve in its centered position. The drain port is blocked. It may be seen that this centered position is achieved by spring loading. This type of directional control valve is generally used like a switch. When the actuator has finished its rapid extension, the valve is moved to its clamping position very rapidly. Likewise, when the part is to be released, the valve is rapidly moved fully over to the retract position.

More information on the various centered configurations of directional control valves may be found in Esposito [3].

7.4   FLAPPER NOZZLE VALVE

There is a trend towards separating an operator from engines and fluid power devices. Sometimes the reason is advanced comfort for the operator and sometimes it is because an operator needs better visibility. A classic example would be the location of the cockpit in an airplane in relation to the location of the engines, pumps, and actuators. For many years there has been a movement towards controlling fluid power valves with electrical signals. As indicated in Section 7.2, solid state electronics now allow solenoids to be equipped with displacement transducers at relatively low cost. Consequently solenoid controlled valves that are capable of proportional control are available to the designer at reasonable cost. An early form of *electrohydraulic* servovalve that still has some application is the flapper nozzle valve coupled to a critical center spool valve. This form of valve is still used by some pump manufacturers to control swash plate displacement, and is also used in aerospace applications.

A typical configuration of such a valve is shown in Figure 7.8. The torsion motor is comprised of an armature with a coil wound on it, a tubular beam that resists armature rotation, and permanent magnets with extended pole pieces. The torsion motor is a specialized form of direct current electric motor that produces a very small angular displacement in response to an applied current. The use of the permanent magnets allows motor reversal with current reversal. The tubular beam serves two purposes. First, it

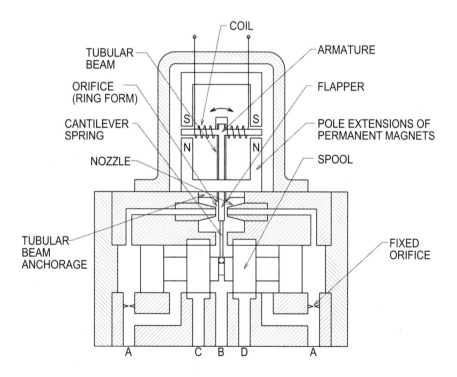

Figure 7.8: Flapper nozzle controlled directional control valve.

allows the armature to rotate through a small angle and second, it seals fluid from the torsion motor.

As with the previous valves discussed in this chapter, the treatment of the flapper nozzle valve will be qualitative. A very extensive treatment of this valve is given by Merritt [2].

As indicated in Section 7.2, a problem with high capacity spool valves is that flow forces are larger than can reasonably be supplied by solenoid actuators. This is even more true for the torsion motor which has a very low force output. Fortunately large force magnification is easily achieved. A small flat plate, the flapper is attached to the torsion motor armature and this plate is centered between two circular, sharp edged ports. These ports are commonly called nozzles, but it should be realized that the bore of the nozzle is parallel. The nozzle effect is achieved by the ring orifice formed between the nozzle and the flapper. At null, the clearance between a port and a plate is less than 0.13 mm. Because of the small clearances and the small angular rotation of the flapper, the motion between the two ports may be considered linear. The flapper and the port form an orifice fairly similar to the orifice in a spool valve. The area of the ring orifice is proportional to the space between the flapper and the nozzle.

It will be observed from Figure 7.8, that there are two orifices in series between the supply pressure and the drain pressure. One orifice is fixed and the other variable. As the nozzle orifice varies in area, the fluid in the volume between the two orifices varies in pressure. As the flapper moves to and fro between the two fixed nozzles, the pressure at one port will rise and the other will fall. Thus giving rise to a pressure differential. The pressure differential operates on the spool treated as an actuator, and the spool moves.

As an example, suppose the flapper moves to the right in Figure 7.8. The ring nozzle to the right will be reduced in area and the pressure will rise as the pressure in the left nozzle drops. The spool will be driven to the left by this pressure differential. The spool now encounters two resisting springs. One is the cantilever spring on the end of the flapper and the other is the effective spring from flow forces, thus the spool will move towards a position of equilibrium where a given value of current gives a certain valve opening. Manufacturers, e.g., [4], claim that for given supply pressure, the spool displacement is proportional to the coil current.

Obviously the flapper nozzle valve is very complex, a designer considering the use of such a valve should obtain information from a manufacturer. Such literature will contain information on working pressures, coil currents, and flow rates [4]. Flapper nozzle valves are generally only used in situations where speed of response is important. Manufacturer's literature will include response information in the form of amplitude vs. frequency re-

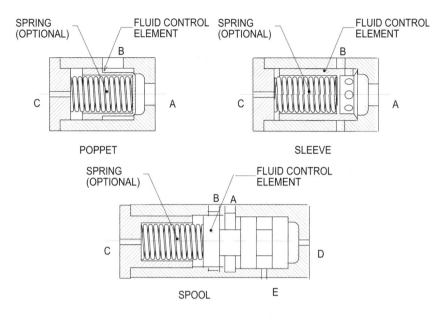

Figure 7.9: Common fluid control elements for valves.

sponse charts.

7.5 FLOW CONTROL ELEMENTS

Before describing valves that have been developed to perform specific tasks, it may be useful to examine some basic aspects of most commonly encountered valves. The fluid control element of most valves takes one of the three forms, poppet, sleeve, or spool shown in Figure 7.9. Used as they have been drawn, all the valves shown in the figure function as pressure relief valves. That is, they pass flow in one direction only and they require a pressure difference across the valve to overcome spring pressure and allow flow.

The poppet valve is probably the least expensive valve because it has the simplest machining requirements. The sleeve valve is also a very common geometry. This form is often incorporated in cartridge valves. The poppet and sleeve geometries are essentially interchangeable functionally for this present discussion and may be examined together. Observe that the poppet and sleeve geometries have three ports, A, B, and C. There is also a spring that has been labeled *optional*. Valves that do not have springs are quite rare, probably only check valves and shuttle valves. Most valves are required to vary their response according to the magnitude of some pressure, a spring is an easy way of achieving that. In the valves shown in Figure 7.9, port A is an entry port and port B a discharge port. Port C is a pressure sensing port and the flow through a sensing port is small to essentially nil. On the other hand, the sensing port may have a vital function. You will observe that none of the valves shown has any form of elastomeric sealing element. Virtually all valves used in fluid power operations leak. The magnitude of the leak is kept small by maintaining very high surface finishes and small clearances. Configurations with seals have generally not been developed because fluid power systems are often required to operate fast and all are required to operate repeatably. Presently no one has developed an elastomeric seal that can perform with an acceptably low static coefficient of friction. If a small amount of fluid leaks through a fluid control element, then only viscous friction will exist and uncertainty in element displacement will be very small.

Most valves perform their desired task by balancing a force developed from a spring with a force developed by differential pressure across the valve. Suppose a poppet valve like that shown in the figure were to have port C plugged. Initially the valve would open if the pressure at port A were sufficiently large, but as oil leaked through into the spring cavity ultimately the pressure pressure differential across the fluid control element would drop to zero. This is a condition of *hydraulic lock*. No matter how high a pressure were applied to port A, the valve would remain closed. Thus the spring cavity must always be drained to an oil volume that is at a lower pressure than that at port A. In many instances, port C will be drained to a reservoir at atmospheric pressure.

In poppet and sleeve valves, one pressure in the pair providing the differential pressure across the flow control element is that of the inlet stream at port A. There are applications where the valve is required to develop its differential pressure from some other fluid stream than that entering the valve. This is an application where a spool valve may provide the simplest solution. The spool valve is a more complicated valve. Note that there are five ports. Ports A, B, and C function as in the poppet and spool valve as inlet, discharge, and pressure sensing. Note, however, that the pressure in port A does not produce a net force on the fluid control element. The spool valve can be made to act as the other two valves by connecting ports A and D. It is the difference between the pressures at ports D and C that cause the pressure differential across the fluid control element of this form of spool valve. Port E is provided so hydraulic lock is avoided. Any inevitable leakage from ports A or D can be drained from port E. The geometry of commercial spool valves may not be exactly as shown, but there will be some provision to avoid hydraulic lock. As will be shown later, a sequence valve may use a spool as a fluid control element (Figure 7.17).

7.6   RELIEF VALVES

Relief valves come in two forms. In the direct acting type, the pressure to be relieved acts directly on the fluid regulating element. This type is commonly employed in systems with relatively low flow rates. Where high flow rates must be passed, a pilot operated type is commonly employed. In this type the pressure to be controlled acts on a pilot. As soon as this pilot allows flow, the pressure difference across the main regulating element becomes large enough to provide a force that causes the valve to open rapidly because the spring controlling the main element is light.

A relief valve may be installed wherever there is a need to protect a device from excessive pressure (Figure 7.10). Probably the most common application would be protecting a pump against an excessive pressure rise when the motion of an actuator or motor becomes blocked. Although the load may be abnormal, actuator motions are often blocked deliberately, for example in a clamping application. Another feature of the circuit shown is the selection of a tandem center directional control valve. Although the system would work with a closed center valve, this valve would be wasteful of energy because the relief valve would open whenever the valve was in the centered position. Should a closed center valve be required for other reasons, the pump should be equipped with a device that changes the displacement to zero at some set pressure. It will be shown later that an unloading valve is an alternative strategy for protecting a pump. Such a valve allows pump to discharge across a low pressure drop.

7.6.1   Direct Acting Type

A direct acting relief valve is shown in Figure 7.11 The moving component is usually a poppet, but other geometries such as a ball or a spool may be encountered. During normal system operation, the regulating element is held in a fluid blocking position by a spring. If the system pressure reaches a preset value, the force on the regulating element will equal the spring preload force. This is called the cracking pressure. Further increase in pressure will cause the regulating element to move from its fluid blocking position and will allow fluid to flow to the system reservoir. Some form of screw adjustment is commonly provided, so the spring preload can be varied. Thus a given relief valve may cover a range of pressure depending on the characteristics of the spring. Because fluid power systems typically operate at high pressures, stiff springs are required. Thus the range of

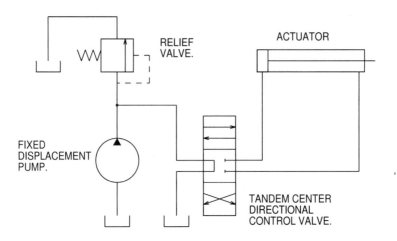

Figure 7.10: Pump protection with a relief valve.

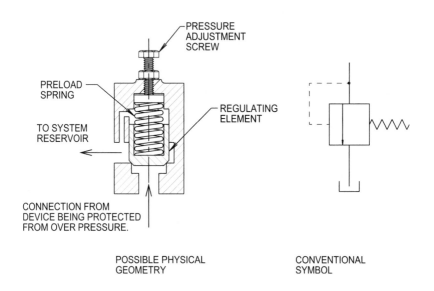

Figure 7.11: Direct acting relief valve.

pressure than can be handled by any given valve spring is limited.

The direct acting relief valve has the disadvantage that pressure rises as the flow though the valve increases [3]. For this reason, this simple form of relief valve is only used for small flow rates.

7.6.2 Pilot Operated Type

A pilot operated relief valve is shown in Figures 7.12 and 7.13 The pilot element can be quite small because this pilot is only required to pass a small flow. During the normal closed condition of the valve, there is no flow through the orifice in the main regulating element, so there is no pressure difference across the main element. Thus the main element can be held in a closed position by a light spring.

As the pressure in chamber A increases, the pressure downstream of the orifice matches this pressure until the pilot valve opens and there is flow to the system reservoir. Now that there is flow through the orifice in the main regulating element, a pressure difference will be developed across the orifice. As indicated earlier, the spring controlling the main regulating element is light, so only a small pressure difference is required to move the main regulating element into its open position. With flow through the valve, there must be sufficient pressure in chamber B to keep the pilot valve open. Thus the pressure in chamber A must exceed the pressure in chamber B by a small amount. The amount being the pressure drop across the orifice in the main regulating element.

Because the main element is controlled by a light spring, the operating pressure is only a small amount above the pressure at which the pilot valve cracked. Consequently the pressure required to open a pilot operated valve is much less dependent on the flow through the valve than is the situation with a direct acting valve.

Although it is necessary to know how a component is constructed physically in order to conduct a dynamic analysis of the component, such drawings would not be satisfactory when representing a circuit. The drawing in Figure 7.13 presents the pilot operated relief valve in schematic form.

7.7 UNLOADING VALVE

A relief valve serves the vital function of providing a path for oil from a pump when the pressure rises to some preset and possibly dangerous level. The disadvantage of a relief valve is that the pressure drop across the valve is large and there will be significant conversion of pressure energy to heat energy. The goal of a circuit designer should be to limit relief valve

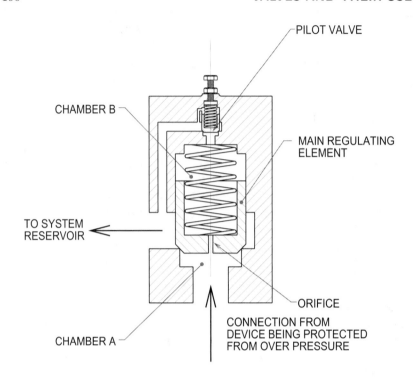

Figure 7.12: Pilot operated relief valve, physical form.

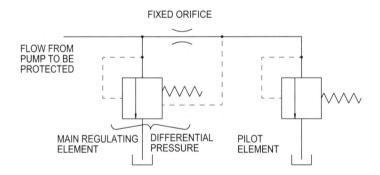

Figure 7.13: Pilot operated relief valve, schematic form.

operations to emergencies. Where possible pump flows should be reduced to zero as pressures rise to limiting values or the relief valve should be replaced by an unloading valve.

An unloading valve is shown in Figure 7.14. The flow from the pump enters at port A and passes through the check valve to the actuator connected at port B. Some fluid power actuators encounter stops as a regular part of their operation. An example would be a cylinder used to clamp an item during a machining operation. The desired function of such a circuit would be to actuate the cylinder until the item is held and the pressure in the system rises to some value considered adequate for the clamping force. At this juncture, no work is being done by the cylinder, so ideally no power should be delivered to the pump supplying the cylinder. The desired pressure is sensed by the feedback line downstream of the check valve (port B). The unloading valve now opens and flow from the pump can pass through the valve to the reservoir across a low pressure. During this low pressure and low power diversion of the pump flow, the pressure in the cylinder is maintained at the desired clamping level by the check valve.

The previous discussion gives no indication of how the pressure to the cylinder can be lowered to allow for retraction. A simple complete circuit for some form of clamp is shown in Figure 7.15. The four way valve now provides the means of lowering pressure downstream of the check valve in the unloading valve. As the four way valve is moved from a clamping position to a release position, the high pressure side of the cylinder is connected to the reservoir allowing the check valve to open. The feedback pressure to the unloading valve also drops and pump flow switches from passing to the reservoir through the unloading valve to passing through the check valve to the rod side of the cylinder. The unloading valve comes into action again as the cylinder reaches its travel limit in the release direction. The pressure in the feedback line will rise again and the unloading valve will open diverting pump flow to the reservoir at low pressure.

7.8 PRESSURE REDUCING VALVE

In some situations, two or more pressures are required in different parts of a circuit, but only one pump is specified to keep costs down. In a pressure reducing valve the pressure sensing line is connected to the low pressure side and serves to close the pressure control element. As a consequence, the spring in the valve serves to open the flow control element. In this situation, the flow control element would open when the pressure in the secondary circuit dropped below the desired value. This change in pressure sensing is not sufficient alone because leakage flow past the flow control element could cause the pressure in the secondary circuit to gradually rise to the

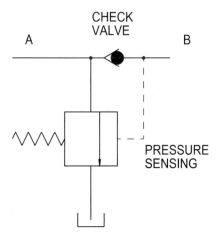

Figure 7.14: Unloading valve.

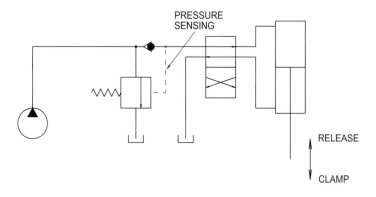

Figure 7.15: Unloading valve used in a clamping application.

pressure in the primary circuit during non operation of the secondary circuit. Although this pressure rise could be controlled by a relief valve in the secondary circuit, a better solution is to provide a small leakage flow from the secondary circuit to the reservoir using a small orifice. The pressure reducing valve showing the necessary features is presented in Figure 7.16.

7.9   PRESSURE SEQUENCING VALVE

It is often necessary for two actions to take place sequentially. For example, in a machining operation the work piece may be clamped by one actuator and after the clamping action is completed a drill head may be moved by another actuator. At the end of clamping, the actuator extension is blocked and the pressure will rise. This pressure rise can be sensed by a sequencing valve and the flow diverted to a secondary circuit. A sequencing valve is a special application of a direct action relief valve (Figure 7.17). The pressure differential across the fluid control element is between the primary circuit pressure and the reservoir. It is important to note that the secondary circuit pressure has minimal effect on the operation of the valve. The check valve is normally included in the body of a pressure sequence valve to allow the

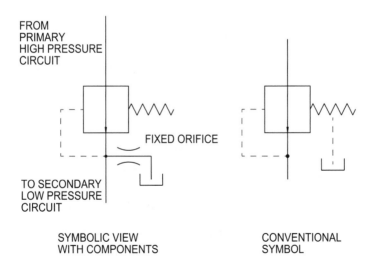

FROM
PRIMARY
HIGH PRESSURE
CIRCUIT

FIXED ORIFICE

TO SECONDARY
LOW PRESSURE
CIRCUIT

SYMBOLIC VIEW
WITH COMPONENTS

CONVENTIONAL
SYMBOL

Figure 7.16: Pressure reducing valve.

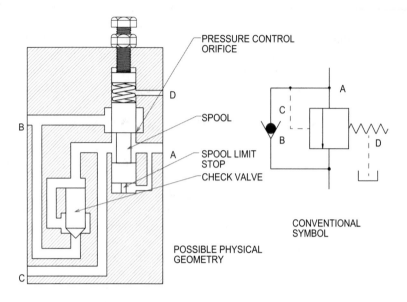

Figure 7.17: Pressure sequencing valve.

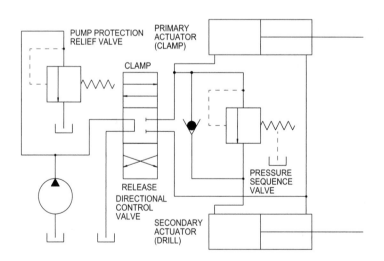

Figure 7.18: Machining operation using pressure sequencing valve.

secondary actuator to retract. Port A is connected to the supply pressure, port B connects to the primary circuit, port C connects to the secondary circuit, and port D drains the spring cavity.

Consider the circuit shown in Figure 7.18. This is a basic circuit for a machining operation that would first clamp a part and then advance a drill to bore a hole. As the actuator in the primary circuit receives fluid from the pump, it is assumed that the pressure is below the set pressure of the sequencing valve. The main fluid control element remains closed. When the extension of the primary cylinder is blocked, the pressure rises and the fluid control element can open allowing fluid to flow into the secondary actuator. When the secondary actuator motion is blocked, the sequence valve remains open and the pressures in the primary and secondary circuits become the same. This pressure is set by a relief valve in the pump circuit. When the clamping pressure is released, the main fluid control element closes. When the directional control valve is moved to release the part, the pressure sequence valve remains closed, but fluid can pass through the check valve and to the directional control valve.

A final comment may be desirable. In the circuit shown in Figure 7.18, the order of release may not be predictable. The actuator that will move first will be the one needing the lower pressure. It may well be desirable to introduce extra components to ensure that the drill is fully retracted before the clamping force on the item being machined is released. For example, a second sequencing valve could be installed in the line from the cap side of the clamping actuator. Thus the clamping actuator would hold the workpiece firmly until the drill was fully retracted and its actuator moved against its stop.

7.10 COUNTERBALANCE VALVE

A counterbalance valve is another variation on the piloted relief valve. It is used to ensure smooth motion of loads that are caused by an external, unidirectional force. This force is commonly caused by gravity. If an actuator is controlled directly by a directional control valve, then lowering a load would begin with a jerk because the pressure in the side supporting the load at rest would suddenly drop to reservoir pressure. When a counterbalance valve is installed, the pressure in the load side of the actuator is always sustained whether the load is being lifted or lowered. The valve itself is shown in Figure 7.19 and a typical application circuit is shown in Figure 7.20. As with a standard relief valve, the frontside of the main fluid control element senses the pressure upstream of the valve. Differential pressure sensing across the element is achieved by sensing the downstream

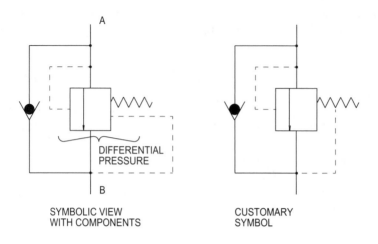

Figure 7.19: Counterbalance valve.

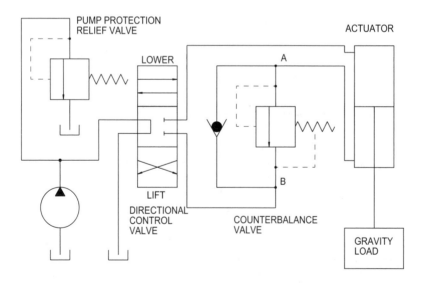

Figure 7.20: Gravity load support with counterbalance valve.

pressure on the backside of the main fluid control element. In addition a check valve is supplied within the valve to allow the valve to accommodate upward movement of the load.

Consider the application circuit shown in Figure 7.20. If the direction control valve is moved to lower the load, fluid under pressure flows from the pump into the cap end of the actuator. Fluid expelled from the rod end passes into port A of the counterbalance valve. In this position of the directional control valve, the pressure on port B is reservoir pressure, so the pressure differential across the main fluid control element is large. The force from the spring operating on the main flow control element is set some amount higher than the force generated by this pressure differential caused by the static load. The valve remains closed until more fluid enters the cap side of the actuator and tries to expel fluid from the rod side. Because the valve is closed, the pressure at port A rises until the spring load is overcome. Oil can now pass through the valve to the reservoir, but the load is never in an unsupported condition, so the lowering starts without a jerk.

The directional control valve is now returned to the centered position. Pressure at port A falls slightly to the value necessary to support the static load. The spring force on the main flow control element now exceeds that caused by the pressure differential across the element and the counterbalance valve closes. The closed valve blocks flow from the rod side of the actuator and the load remains fixed in place.

Now consider the directional control valve being moved to lift the load. Oil from the pump appears at port B. Initially, there is no open path for the oil, so the pressure rises. The pressure differential across the main flow control element drops allowing the spring to maintain the element in its closed position. Flow to the rod side of the actuator will bypass the main flow control element and pass through the check valve. Fluid expelled from the cap side of the actuator passes freely through the directional control valve to the reservoir.

The directional control valve is now returned to the center position. The check valve returns to its seat and the counterbalance valve remains closed because the pressures at port A and B are the same. As before, the load remains stationary.

There are occasions when a gravity load on an actuator can reverse. One example would be an actuator supporting some form of mast that can move either side of a vertical position. A single counterbalance ceases to work if the load on the valve is reversed. In such a situation, two counterbalance valves must be provided operating in reverse directions. Which valve operates is determined by a shuttle valve that switches oil to the correct valve [5].

7.11   FLOW REGULATOR VALVE

In some circuits, it is important to ensure that the flow rate to a device is independent of the pressure in the device circuit. Before discussing a method of achieving this goal, it should be mentioned that exact flow control is not possible with a flow regulator valve. Esposito [3], however, indicates that flow rate can be controlled to an error of less than 5%. If more accurate control of actuator response is necessary, then some form of feedback and servo control would be necessary.

It should be obvious that some form of flow control can easily be obtained by placing a restriction in the supply line to the actuator. If limiting speed is all that is required, this form of flow controller may be quite adequate. In many instances, however, the designer would like flow rate to be largely independent of circuit pressure. If the pressure differential across the orifice can be kept constant then the flow through the orifice will be independent of the pressure in the circuit being supplied. The pressure compensated flow regulator valve shown in Figure 7.21 is a means of achieving constant pressure differential across an orifice.

As with the pressure sequence valve discussed in Section 7.9, a spool is used for the flow control element because this valve needs to sense some pressure other than that in the supply line. Initially consider the pressure at port B to be atmospheric, that is the actuator is at rest, retracted, and not under load. The pressure at port A is at some fraction of system pressure, probably provided by the main system pump via a directional control valve. Because p_2 is atmospheric, the spool will be displaced to hold the pressure control orifice open and there will be flow through this orifice. The internal pressure in the valve, p_1, will rise and cause the pressure control orifice to close. There will also be flow through the user variable orifice to port B and the secondary circuit. An equilibrium will be established so the flow though the valve reaches the desired value. The differential pressure across the spool, $p_1 - p_2$, causes a force that matches the spring force and allows the pressure control orifice to open just enough to match the flow out of port B.

Notice that the load on the actuator can change causing a change in p_2, but the spool will adjust its position so p_1 changes and sustains a constant value of pressure differential. The pressure differential across the spool will remain constant, but the pressure control orifice opening will alter slightly to maintain the flow through this orifice at the same value as the flow through the user orifice. The pressure difference across the pressure control orifice will not remain constant because the downstream pressure p_1 will vary as p_2 varies.

Now suppose the actuator connected to the flow regulator valve reaches

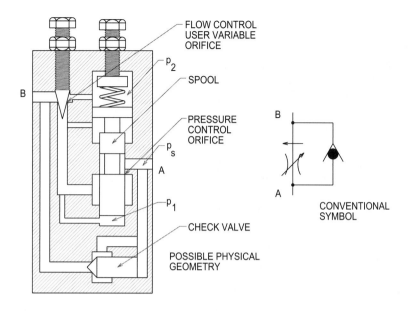

Figure 7.21: Flow regulator valve.

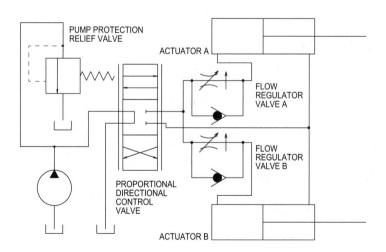

Figure 7.22: Flow regulator valves in circuit used to coordinate two actuators.

200 VALVES AND THEIR USES

a blocked position. The pressure p_2 will rise until it reaches p_1, so the pressure control orifice will open because the pressure differential is zero. Both p_2 and p_1 will very rapidly reach the supply circuit pressure p_s. The pressure control orifice will remain at maximum opening. Consequently there is no need to provide a relief valve in the actuator circuit because the supply circuit relief valve will protect the actuator circuit.

If the directional valve is changed to the retract position, p_2 would exceed p_1 because of reverse flow through the user orifice. The resulting pressure differential, $p_2 - p_1$, would cause the pressure control orifice to open, but the pressure differential across the user orifice is not regulated for reverse flow. A flow regulator valve is commonly equipped with a check valve to allow essentially unrestricted flow during retraction.

A simple application of flow regulator valves is shown in Figure 7.22. Suppose actuators A and B are of different capacity, but both must reach specific operating positions at nearly the same time. As indicated earlier, flows can only be coordinated to about 5%. Both actuators are supplied through pressure compensated, flow regulator valves, so the flows to each actuator may be regulated by the user to satisfy the speed requirements. This is done manually by adjusting the flow control user variable orifices. Once set, the flows will coordinate as required even if the pressures experienced by the actuators vary independently.

PROBLEMS

7.1 Consider the remarks at the end of the section discussing the pressure sequence valve that are associated with tool and clamp withdrawal in Figure 7.18. Draw an improved circuit that will ensure that the drill actuator is fully retracted before the clamp actuator begins to release. Write a complete sequence of operations specifying the fluid flows and how the valves open and close.

7.2 A flow control valve is designed to control the speed of the hydraulic motor as shown in the figure.

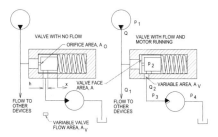

Determine the variable flow area, A_V, the pressure downstream of the valve fixed orifice, p_2, The valve displacement, x, and the valve spring preload, F, for the given motor operating conditions.

Characteristics of a system using a flow control valve

Characteristic	Size	Units
Length, h	7.8	mm
Valve area gradient for flow area A_V, w	1.25	mm^2/mm
Valve flow coefficients	0.6	
Spring rate	57.0E+03	N/m
Fixed orifice flow area, A_O	4.9	mm^2
Valve face area, A	125	mm^2
Oil mass density, ρ	840	$N \cdot s^2/m^4$
Motor displacement, D_m	40.0	cm^3/rev
Motor torque	60.0	$N \cdot m$
Motor speed, n	350	rpm
System pressure, p_1	14.5	MPa
Return pressure, p_4	1.0	MPa
Motor volumetric efficiency, η_V	96.0	%
Motor torque efficiency, η_m	97.0	%

7.3 A flow control valve is designed to control the velocity of the actuator as shown in the figure.

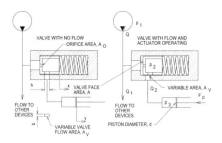

Determine the cylinder force, F_p, and the cylinder piston velocity, $\dot{y}$, that is developed in the system when cylinder flow, Q_2, is equal to 80% of the pump flow Q.

Characteristics of a system using a flow control valve

Characteristic	Size	Units
Length of flow area A_V, h	0.24	in.
Width of flow area A_V, b	0.059	in.
Valve flow coefficients, C_d	0.6	
Spring rate, k	375.0	$\text{lb}_f/\text{in.}$
Fixed orifice flow area, A_O	7.8E$-$3	in.^2
Valve face area, A	0.2	in.^2
Piston diameter, d	1.5	in.
Spring preload, F	57.0	lb_f
Oil mass density, ρ	0.08E$-$3	$\text{lb}_f \cdot \text{s}^2/\text{in.}^4$
System pressure, p_1	2025	$\text{lb}_f/\text{in.}^2$
Pump flow, Q	5.0	gpm

7.4 A flow control valve is designed to control the velocity of the actuator as shown in the figure.

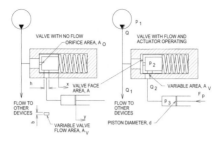

Determine the cylinder force, F_p, and the cylinder piston velocity, $\dot{y}$, that is developed in the system when cylinder flow, Q_2, is equal to 80% of the pump flow Q.

Characteristics of a system using a flow control valve

Characteristic	Size	Units
Length of flow area A_V, h	6.0	mm
Width of flow area A_V, b	1.5	mm
Valve flow coefficients, C_d	0.6	
Spring rate, k	65.7	kN/m
Fixed orifice flow area, A_O	5.0	mm^2
Valve face area, A	130	mm^2
Piston diameter, d	38.0	mm
Spring preload, F	255	N
Oil mass density, ρ	830	N·s^2/m^4
System pressure, p_1	13.97	MPa
Pump flow, Q	0.32	L/s

7.5 A pilot operated cartridge valve is being used as a pressure regulating valve. The system pressure, p_1, will cause the pilot valve G to open. The main flow control sleeve D will move towards the left because of the pressure differential caused by the flow, Q_E, through the pilot supply orifice, E.

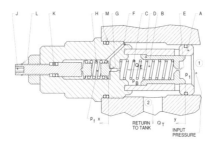

The features of the valve are:

A. Floating body
C. Minimum pressure bias spring
E. Pilot supply orifice
G. Pilot closure
J. Adjusting screw
L. Overpressure limiter

B. Control chamber
D. Main flow control sleeve
F. Pilot return to tank passage
H. Pilot spring
K. Seal for adjusting screw
M. Thread and static seal assembly

Characteristics of a cartridge valve

Characteristic	Size	Units
Flow coefficient for all orifices, C_d	0.61	
Pilot supply orifice diameter, d_E	0.025	in.
Sleeve outside diameter, d_D	0.475	in.
Rate for spring C, k_C	325	$lb_f/in.$
Oil density, ρ	0.08E−3	$lb_f \cdot s^2/in.^4$
Pressure downstream of pilot closure, p_F	50	$lb_f/in.^2$
Tank pressure, p_T	0	$lb_f/in.^2$

The valve flow area for tank flow is:

$$A_t = 4(-0.1608\text{E}{-3}z + 0.33z^{1.5} - 1.845z^{2.5}) \text{ in.}^2$$

the parameter, z, is the actual distance that the four holes have been opened to flow. The pilot valve has dimensions such that the flow through the pilot valve at a displacement x is $Q_G = 12 \times \sqrt{\Delta p}$ in.^{3}/s.

Determine the flow, Q_E, through the pilot valve when valve motion, x, is equal to 0.002 in. Determine the value of control chamber pressure,

p_B, for this condition. The main valve sleeve, D, has an overlap, U, equal to 0.090 in. before the round holes begin to open to provide an initial tank flow, Q_T, to the tank. Determine the initial force, F_y, that is required in the spring C to provide an initial tank flow, Q_T. Determine the tank flow, Q_T, when the main sleeve has moved the distance $y = U + 0.004$ in.

7.6 A pilot operated cartridge valve is being used as a pressure regulating valve. The system pressure, p_1, will cause the pilot valve G to open. The main flow control sleeve D will move towards the left because of the pressure differential caused by the flow, Q_E, through the pilot supply orifice, E.

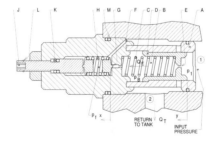

The features of the valve are:

A. Floating body	B. Control chamber
C. Minimum pressure bias spring	D. Main flow control sleeve
E. Pilot supply orifice	F. Pilot return to tank passage
G. Pilot closure	H. Pilot spring
J. Adjusting screw	K. Seal for adjusting screw
L. Overpressure limiter	M. Thread and static seal assembly

Characteristics of a cartridge valve

Characteristic	Size	Units
Flow coefficient for all orifices, C_d	0.61	
Pilot supply orifice diameter, d_E	0.02	in.
Sleeve outside diameter, d_D	0.425	in.
Rate for spring C, k_C	300	$\mathrm{lb_f/in.}$
Oil density, ρ	0.08E$-$3	$\mathrm{lb_f \cdot s^2/in.^4}$
Pressure downstream of pilot closure, p_F	50	$\mathrm{lb_f/in.^2}$
Tank pressure, p_T	0	$\mathrm{lb_f/in.^2}$

The valve flow area for tank flow is:

$$A_t = 6\left[R^2 \arccos\left[\frac{(R-z)}{R}\right] - (R-z)\sqrt{2Rz-z^2}\right] \text{ in.}^2$$

the parameter, z, is the actual distance that the six holes have been opened to flow with a hole diameter $2r = 0.05$ in. The pilot valve has dimensions such that the flow through the pilot valve at a displacement y is $Q_G = 14 \times \sqrt{\Delta p}$ in.^{3}/s.

Determine the flow, Q_E, through the pilot valve when valve motion, x, is equal to 0.0018 in. Determine the value of control chamber pressure, p_B, for this condition. The main valve sleeve, D, has an overlap, U, equal to 0.092 in. before the round holes begin to open to provide an initial tank flow, Q_T, to the tank. Determine the initial force, F_x, that is required in the spring C to provide an initial tank flow, Q_T. Determine the tank flow, Q_T, when the main sleeve has moved the distance $y = U + 0.003$ in.

REFERENCES

1. Wolansky, W. and Akers, A., 1988, *Modern Hydraulics the Basics at Work*, Amalgam Publishing Company, San Diego, CA.

2. Merritt, H. E., 1967, *Hydraulic Control Systems*, John Wiley & Sons, New York, NY.

3. Esposito, A., 1999, *Fluid Power with Applications*, 6th ed., Prentice Hall, Englewood Cliffs, NJ.

4. Moog, Inc., Viewed June 28 2003, "Electrohydraulic valves...A Technical Look", http://www.moog.com/Media/1/technical, Moog, Inc., East Aurora, New York, NY.

5. Pippenger, J. J., 1990, *Hydraulic cartridge valve technology*, Amalgam Publishing Company, Jenks, OK.

8

PUMPS AND MOTORS

8.1 CONFIGURATION OF PUMPS AND MOTORS

Two phenomena may be used for pumping. The first phenomenon employs a chamber that is initially attached to an inlet port. The chamber increases its volume to accommodate fluid forced into it by the prevailing fluid pressure at the inlet. The chamber is then connected to an outlet port and reduced in volume to expel the fluid. The discharge pressure is greater than the inlet pressure. Pumps constructed to operate using this phenomenon are *positive displacement* or *hydrostatic* pumps. If leakage and mechanical strength considerations are ignored, then a positive displacement pump can pump fluid at any speed and over any pressure difference. The second phenomenon employs a spinning rotor that can impart a tangential velocity to the fluid. There is no gross flow of fluid tangential to the rotor, so ideally the pressure that could be achieved by converting the tangential velocity to pressure would be:

$$p = \rho \frac{(\omega_m r))^2}{2} \tag{8.1}$$

Pumps employing this phenomenon are *hydrodynamic* pumps.

Fluid power systems operate at pressures such that only positive displacement pumps provide the combination of size, performance, and efficiency that is acceptable. A hydrodynamic pump can operate with a closed discharge for a short time, i.e., before the fluid overheats. On the other hand, closing the discharge of a positive displacement pump generally invites some form of immediate catastrophic failure, for example shearing of the drive shaft. Consequently, all positive displacement pumps must be protected with relief valves or some form of mechanical device that reduces the positive displacement to zero at some preset pressure.

In this chapter, pumps and motors are generally discussed as if there were no difference between the two functions. This lack of discrimination is generally true and pumps may often be substituted for motors and *vice versa*. It is NOT universally true. For example, a gerotor pump is simpler than a gerotor motor, for reasons that will be explained when discussing that specific geometry. Manufacturers design pumps and motors for specific purposes and it is always advisable to consult with the manufacturer should the intended role of a piece of equipment be changed.

The common patterns of positive displacement pumps employed in fluid power systems are displayed in Figure 8.1 to 8.8. It will be observed that all the patterns shown have some means of connecting the expanding chamber to an inlet port and the contracting chamber to a discharge port. Achieving a graceful transition is not trivial. Excess leakage vs. excess pressure rise must be balanced. We shall discuss this topic later for the axial piston pump in Chapter 9. Both pumps and motors are generally ranked in the order:

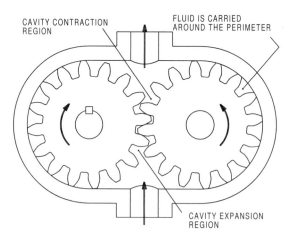

Figure 8.1: Spur gear pump.

SPUR GEAR
INTERNAL GEAR
GEROTOR
VANE
RADIAL PISTON
AXIAL PISTON (SWASH PLATE & BENT AXIS)
SCREW

The initial pumps in the ranking are relatively lower in cost, limited in pressure capability, and lower in efficiency. In addition to pressure capability and efficiency, piston pumps are relatively easily modified to incorporate automatic volume displacement reduction as pressure rises above a set value. Furthermore piston pumps and motors can have their displacement varied during normal operation. This characteristic is made use of in a hydrostatic transmission where motor speed must be varied smoothly (i.e. steplessly) over a wide range.

Spur gear: A spur gear pump is shown in Figure 8.1. The fluid is carried around the perimeter of the gear. The cavity expansion and cavity contraction take place where the gears mesh. It may be observed that the actual volume change is limited to the volume of one inter tooth space. There are practical limitations to the thickness of the gears so the volu-

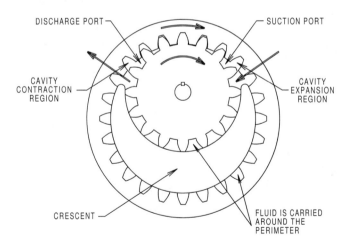

Figure 8.2: Internal gear pump.

metric capacity of spur gear pumps for a given case volume is relatively low.

As with all fluid power pumps or motors, the only elastomeric seals are those around the shaft. The long passage ways in the spur gear pump, for example around the gear perimeter, control leakage quite well. Figure 8.1 has been drawn with realistically shaped teeth. It will be observed that there is significant fluid that is never expelled from the contracting cavity. As discussed in Chapter 2, hydraulic fluids possess elasticity so any dead volume lowers volumetric efficiency as the working pressure increases.

Internal gear: The internal gear pump shown in Figure 8.2 is the second of the variations on gear pumps. For the same number of teeth in the ring gear as the gear of a spur gear pump, the internal gear pump will be more compact. An internal gear pump will also have a slightly smaller dead fluid volume than a spur gear pump because of the tooth meshing geometry.

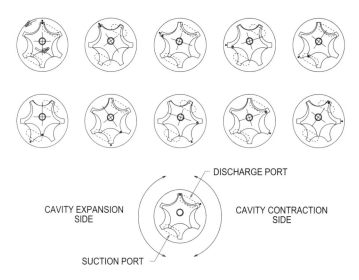

Figure 8.3: Gerotor motor.

Gerotor: The gerotor geometry is a patented form of the internal gear geometry. The word *geometry* is used because the gerotor is one device where the pump and motor may be different.

The gerotor pump is close relative of the internal gear pump. The axes of the spur and ring gears are offset and fixed in their relative locations. The operation and method of fluid transfer is essentially the same as for the internal gear except that the geometry does not require the crescent insert required in the internal gear pump. This form of pump is commonly used as a charge pump for hydrostatic transmissions (Section 10.3) because it has a minimum number of parts

As with other positive displacement geometries, the fixed axis geometry can be used for a pump or a motor. This simple geometry is seldom used for gerotor *motors* because the addition of a few components can alter the displacement per revolution of the device in a major fashion. A special drive shaft in the form of a diminutive dumbbell is added. This shaft has male splines at both ends, but these splines are curved in the plane of the shaft, that is, the splines are barrel shaped. The ring gear is driven around a fixed axis guided by a fixed circular bearing surface in the pump case. The rotor rotates about an axis that is slightly offset from the ring axis (Figure 8.3). In addition, the rotor orbits around the ring gear axis during operation, hence the need for the dumb bell drive shaft.

The manner in which expanding and contracting cavities are formed during rotation can be seen in Figure 8.3 sequence. The gerotor motor in the orbiting form has one major advantage compared to the simple form with fixed axes. The displacement per revolution is much larger than that which can be obtained with the fixed axis geometry for the same rotor and stator size.

An alternative version of the gerotor motor cuts cylindrical slots in the ring gear and places rollers in these slots. This modification reduces friction and allows the units to be used at much lower speeds than most other motor geometries. The motor is called a *geroller* in this configuration. A typical application would be as a direct drive motor for a wheel. The terminology *low speed high torque* is used for such motors.

Vane: Vane pumps are available in unbalanced and balanced geometries. Figure 8.4 shows the balanced configuration. The unbalanced type is similar but consists of a circular cavity offset from the shaft axis. Consequently, there is an unbalanced radial force on the drive shaft resulting from the differences in pressures in this cavity. For pumps used at low pressures, such an unbalanced force is not a problem, but for the pressures employed in fluid power systems, eliminating unbalanced forces is good practice. The springs shown between the inner edges of the vanes and the rotor are optional. When included they ensure that the outer edges of the vanes and the case are in contact at rest.

Radial piston: Radial piston pumps exist in two main geometries. In

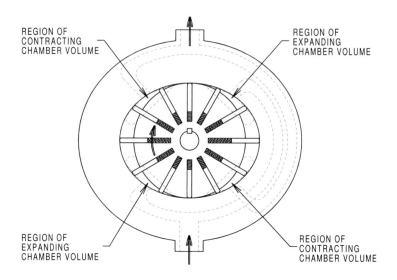

REGION OF
CONTRACTING
CHAMBER VOLUME

REGION OF
EXPANDING
CHAMBER VOLUME

REGION OF
EXPANDING
CHAMBER VOLUME

REGION OF
CONTRACTING
CHAMBER VOLUME

Figure 8.4: Balanced vane pump.

the geometry shown in Figure 8.5, the axis of each piston is fixed in space.

In the other geometry, not shown, the piston axes orbit around the drive shaft axis and piston displacement is caused by the eccentricity of the casing on which the outer ends of the pistons bear. There are no valves and fixed ports are provided in a central cavity surrounding the shaft.

The geometry in Figure 8.5 is shown because it was used by John Deere in many of their tractors. The radial pump shown in Figure 8.5 does not exactly represent a Deere pump because the valve positions have been altered to simplify the presentation. In practice, the valves are aligned with their bores parallel to the drive shaft axis.

The geometry is interesting because it allows stroke control and over-pressure protection. Stroke control is achieved by controlling the pressure in the stroke control cavity. The inward radial projection of the pistons is determined by the equilibrium between the stroke control spring force and the product of stroke control cavity pressure and piston cross section area. Consequently, this geometry of radial piston pump can operate as a controlled variable displacement pump. Facilities may also be provided to raise the stroke control cavity pressure to an upper limiting value such that there is no pumping action if the pump output pressure reaches a preset limiting value. Radial piston pumps commonly employ an even number of

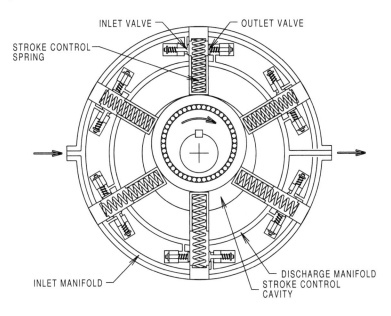

Figure 8.5: Radial piston pump with displacement control.

pistons because boring a pair of cylinders at one setting reduces production cost.

Axial piston, swash plate: In an axial piston pump, a piston is reciprocated by means of motion relative to an inclined plane. Usually the inclined plane is stationary and is called a swash plate. As shown in Figure 8.6, the barrel containing the pistons rotates to give the relative motion between the pistons and the inclined plane. The swash plate angle is limited to 18° for satisfactory mechanical operation.

At top dead center and bottom dead center, the fluid in the cylinder must make a transition from high (or low) to low (or high) pressure. This transition is achieved with the tapered channel, the pressure transition groove. The objectives of this groove are first to ensure that fluid is never trapped in a changing volume that lacks a passage and second to limit communication between the high and low pressure manifolds to a very small passage. The first objective is necessary to avoid excessive pressures developing in the cylinder and the second ensures that efficiency is maintained.

The number and size of grooves that may be observed in a given device may vary. This is another instance where pumps may vary from motors.

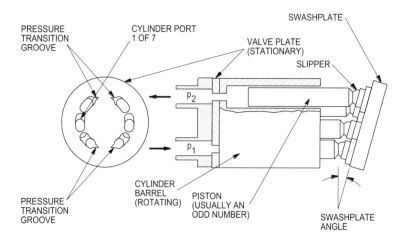

Figure 8.6: Axial piston pump, swash plate type.

A variable displacement pump will generally be constructed to allow the swashplate to move $\pm 18°$ about a null position, thus the top dead center for the pump will switch by $180°$ on the valve plate. In a motor, reversal is usually permitted, so the direction in which a cylinder approaches top dead center will switch. As will be discussed in Chapter 9, the magnitude of the low pressure on the pump or motor will affect the size of the grooves (p. 247).

It will be shown in Chapter 9 that the variation in flow rate from a pump with an odd number of pistons is much less than that from a pump with an even number. In practice, five, seven, nine, or occasionally 11 pistons are used. Variation in flow rate causes pressure variation and this, in turn, causes noise.

Axial piston, bent axis: The pumping elements of an axial piston pump using the bent axis configuration are essentially the same as those used for an axial piston pump with a swash plate. The major difference results from the means of moving the inclined plane past the pistons. The inclined plane is attached to the pump drive shaft and is normal to that shaft at all times (Figure 8.7). The inclination is achieved by setting the cylinder block axis and the drive shaft axis at an angle (hence the name bent axis). This angle may be up to $40°$. The cylinder block and drive shaft are constrained to rotate together by an internal drive shaft. The ends of this shaft are provided with joints that will allow rotary motion as

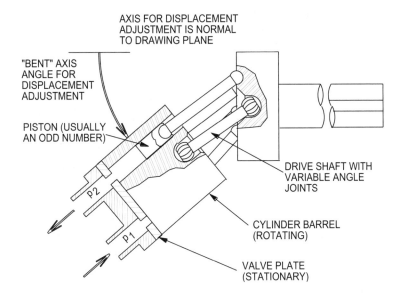

Figure 8.7: Axial piston pump, bent axis type.

the bent axis angle is changed.

Reciprocating motion between the drive shaft plate and the cylinder barrel is transmitted by push rods with ball ends. These ball ends are necessary because the push rods do not remain parallel during rotation.

Screw: The screw pump has been saved until last because of its position in the hierarchy of pressure and cost. In fact, it has much in common with gear pumps. As with these geometries, the screw pump has a fixed displacement and a packet of fluid is moved through the pump against the case from the cavity expansion region to the cavity contraction region.

On the other hand, the screw pump has two beneficial features. Inspection of the specific geometry presented in Figure 8.8 shows that at the discharge there are three contraction regions discharging simultaneously. This means that the screw pump has a very small variation of flow rate with rotation. Consequently, systems using screw pumps have the potential for being much quieter than systems using most other forms of pump.

The second feature is that screw pumps may be designed with long sealing lengths to control leakage and this feature allows the screw pump to develop pressures in excess of most other geometries. The screw pump

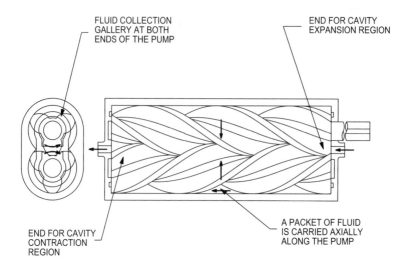

FLUID COLLECTION
GALLERY AT BOTH
ENDS OF THE PUMP

END FOR CAVITY
EXPANSION REGION

END FOR CAVITY
CONTRACTION
REGION

A PACKET OF FLUID
IS CARRIED AXIALLY
ALONG THE PUMP

Figure 8.8: Screw pump.

may be found in a wide variety of configurations. The form presented in
Figure 8.8 is designed for a high flow rate, but only modest pressure. An
alternative configuration uses two square or acme thread form screws with
the threads meshed. In this form, there might be 15 to 20 threads along
the length of a rotor. Here the discharge per revolution of the pump would
be relatively small, but the pressure capability very large.

Two other features that may be encountered are timing gears and more
than two rotors. The geometry shown in Figure 8.8 uses involute teeth
and does not require timing gears. High pressure pumps using square or
similar screws will require timing gears to maintain synchronous rotation.
The other feature is use of multiple rotors. An additional slave rotor can
be installed parallel to the main drive rotor. The extra rotors improve the
displacement of the pump and may cost less than increasing the rotor size.
The screw pump is discussed in more detail in Karassik et al. [1].

8.2 PUMP AND MOTOR ANALYSIS

The form of analysis that follows is useful for characterizing existing equipment, but has limited application for the designer of new equipment. In
spite of this shortcoming, we believe that the analysis will help a system
designer understand the form of curves shown on manufacturer's specifica-

tion sheets. The analysis that will be developed follows that of Merritt [2]. Let us first examine an ideal motor. In such a motor, the following factors would be absent:

LEAKAGE

FRICTION

FLUID COMPRESSIBILITY

CAVITATION

The power output of an ideal motor is:

$$P = T_g \omega_m \tag{8.2}$$

Also for an ideal motor, it is possible to express the power in terms of the flow through the motor and the pressure drop across the motor:

$$P = Q_L p_L \tag{8.3}$$

Note incidentally that both equations imply using consistent units, for example SI units. The two power expressions can be equated to eliminate P yielding:

$$T_g \omega_m = p_L Q_L$$

or:

$$T_g = \frac{Q_L}{\omega_m} p_L$$

The quantity Q_L/ω_m has dimensions volume/radian. This quantity is a characteristic of a specific motor or pump and is called the displacement, D_m or D_p. Thus a simple, yet useful, expression relating the torque expected from an ideal motor is:

$$T_g = D_m p_L \tag{8.4}$$

Observe, however, that manufacturers usually will provide the information in in.^{3}/rev or cm^3/rev rather than volume/radian, which is needed for application in Equation 8.4. Any real pump or motor will have friction and leakage so the expressions that should be used are:

$$T_g = \eta_m D_m p_L \qquad 0.85 < \eta_m < 0.96 \quad \text{approximately} \tag{8.5}$$

and:

$$T_g = \frac{D_p p_L}{\eta_p} \qquad 0.85 < \eta_p < 0.96 \quad \text{approximately} \tag{8.6}$$

8.2.1 Example: Drive for a Hoist

A hoist is required to lift a mass of 1000 kg. The cable is wound on a drum 0.15 m in diameter. The drum is driven by a hydraulic motor through a 5:1 reduction gearbox. The system pressure is 10 Mpa (1400 lb/in.^{2}) and the motor discharges to atmospheric pressure. Find the ideal motor displacement required. Make a recommendation for the motor displacement that should actually be provided.

Torque on the drum shaft:

$$1000 \text{ kg} \times 9.81 \, \frac{\text{m}}{\text{s}^2} \times \frac{0.15}{2} \text{ m} \;=\; 735.75 \, \frac{\text{kg} \cdot \text{m}^2}{\text{s}^2}$$

$$= \;\; 735.75 \text{ N} \cdot \text{m}$$

Torque at the motor:

$$\frac{735.75}{5} = 147.15 \text{ N} \cdot \text{m}$$

Using:

$$T_g = D_m p_L$$

Then:

$$D_m \;=\; \frac{T_m}{p_L} = \frac{147.15 \text{ N} \cdot \text{m}}{(10\text{E}+6 - 0) \text{ N/m}^2}$$

$$= \;\; 14.72\text{E}{-}6 \text{ m}^3/\text{rad}$$

Being conservative and selecting a worst case value for the motor efficiency of 85%, then the motor ought to have a displacement of:

$$\frac{14.72\text{E}{-}6}{0.85} = 17.32 \text{ m}^3/\text{rad}$$

Expressing this result in units that you would see in a manufacturer's catalogue:

$$17.32\text{E}{-}6 \times 2\pi(100 \text{ cm/m})^3 \;=\; 108 \text{ cm}^3/\text{rev}$$

$$= \;\; \frac{108 \text{ cm}^3/\text{rev}}{(2.54 \text{ cm/in.})^3}$$

$$= \;\; 6.64 \text{ in.}^3/\text{rev}$$

8.3 LEAKAGE

Although seven different geometries (SPUR GEAR to SCREW) were listed earlier, we may analyze leakage with a generic model based on an axial piston geometry.

For any positive displacement pump or motor, there is an inlet chamber that is increasing in volume and a discharge chamber that is decreasing. There may be leakage directly from the high pressure discharge chamber to the low pressure inlet chamber. There may also be discharge from both chambers to the drain. The drain is some volume of the pump maintained at atmospheric pressure. Elastomeric seals on fluid power pumps are limited in application. Typically these seals are only used where the shaft passes through the case to the outside. Leakage elsewhere in the device is controlled by very fine machining tolerances. Although this approach minimizes sliding friction, it does mean that leakage is always present. It should be noted that leakage is not totally detrimental. Leakage provides oil for lubrication.

The term *drain* is used to indicate a volume at atmospheric pressure separate from the inlet and discharge volumes. Most pumps and motors have such volumes and these volumes are provided with a port so the fluid can be drained to a reservoir. The drain volume is ported to the volume inboard of the shaft seal because there should be minimum pressure difference across the shaft seal to limit leakage.

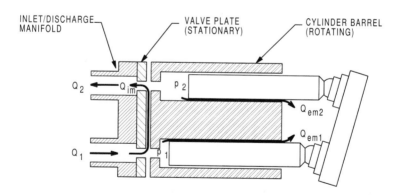

Figure 8.9: Basic leakage paths for a motor.

Figure 8.9 displays the basic leakage paths for a motor. You should observe that the paths shown are greatly exaggerated in size. Clearances in a real device would be in the order of 5 μm (0.0002 in.). Because the clearances are so small, leakage flow is treated as laminar so:

$$Q \propto \Delta p$$

Thus for a motor we have:

Inlet to outlet leakage	Q_{im}	$=$	$C_{im}(p_1 - p_2)$
Inlet chamber to drain	Q_{em1}	$=$	$C_{em}(p_1 - p_0)$
Outlet chamber to drain	Q_{em2}	$=$	$C_{em}(p_2 - p_0)$

Note that p_0 is usually atmospheric pressure.

In an open circuit system, i.e., where a pump draws from a reservoir at atmospheric pressure and a motor discharges to a similar reservoir, p_2 may essentially be at atmospheric pressure. On the other hand, in a hydrostatic transmission the suction line will be maintained at 2 to 3 Mpa (280 to 430 lb/in.^{2}). The elevated suction line pressure is provided by a charge pump (Section 10.3) and this pressure eliminates any possibility of suction line cavitation. Apply the principle of continuity to the inlet and outlet chambers:

$$
\begin{aligned}
Q_1 \quad = \quad & \text{Ideal motor rate of flow} \\
& +\text{Leakage to outlet} \\
& +\text{Leakage to drain}
\end{aligned}
$$

$$Q_1 = D_m\omega + C_{im}(p_1 - p_2) - C_{em}(p_1 - p_0) \qquad (8.7)$$

Similarly for the suction chamber:

$$
\begin{aligned}
Q_2 \quad = \quad & \text{Ideal motor rate of flow} \\
& +\text{Leakage from inlet} \\
& +\text{Leakage to drain}
\end{aligned}
$$

$$Q_2 = D_m\omega + C_{im}(p_1 - p_2) - C_{em}(p_2 - p_0) \qquad (8.8)$$

It is convenient to introduce a *load flow* variable defined as:

$$Q_L = (Q_1 + Q_2)/2 \qquad (8.9)$$

Thus:

$$Q_L = D_m\omega + \left(C_{im} + \frac{C_{em}}{2} \right)(p_1 - p_2) \qquad (8.10)$$

Now introduce a further variable *load pressure*:

$$p_L = p_1 - p_2 \tag{8.11}$$

Thus:

$$Q_L = D_m\omega + \left(C_{im} + \frac{C_{em}}{2}\right)p_L \tag{8.12}$$

This expression for Q_L is often useful when performing system simulation and a mean flow through the device is required.

The volumetric efficiency of a pump or motor can be specified under any set of operating conditions. When manufacturers quote performance characteristics in technical literature, they need to do so under some baseline condition. It is generally accepted that manufacturer's literature fixes the pump suction or motor discharge condition at zero gauge pressure.

Introduce a relationship between the leakage coefficients, viscosity, and motor displacement:

$$C_s\frac{D_m}{\mu} = C_{im} + C_{em} \tag{8.13}$$

You might wonder why $C_{im} + C_{em}$ and not $C_{im} + C_{em}/2$. When pump or motor volumetric efficiency is being specified, only the flow into a motor or out of a pump is of concern. Hence for a motor under baseline conditions with $p_2 = 0$, the flow into the motor is:

$$
\begin{aligned}
Q_1 &= D_m\omega + (C_{im} + C_{em})p_1 \\
&= D_m\omega + C_s\frac{D_m}{\mu}p_1 \tag{8.14}
\end{aligned}
$$

The component $Q_s = (C_{im}+C_{em})p_1$ is sometimes called slip flow. Examine the dimensions of C_s:

$$
\begin{aligned}
Q_s &= (C_{im} + C_{em})p_1 \\
C_{im} + C_{em} &= \frac{Q_s}{p_1} = \frac{L^3}{T}\cdot\frac{LT^2}{M} = \frac{L^4T}{M}
\end{aligned}
$$

Now examine:

$$\frac{D_m}{\mu} = L^3\cdot\frac{LT}{M} = \frac{L^4T}{M}$$

Thus C_s is dimensionless.

8.3.1 Example: Estimating Pump Performance Coefficient C_s

This example will examine the expected magnitude of C_s using data taken from the specification of a Sauer Sundstrand Series 20 pump. The specification sheet indicates that the volumetric efficiency is 96.5% at 3000 rpm when $p_L = 3000$ lb/in.^{2} (remember that $p_L = p_1$ and $p_2 = 0$). The nominal maximum displacement of this unit is 2.03 in.^{3}/rev. This example will be worked in inch-pound force-second units because this approach will give an opportunity to review conversions between various forms of viscosity units.

Start by defining the volumetric efficiency:

$$\text{VOLUMETRIC EFFICIENCY (PUMP)} \quad = \quad \frac{\text{ACTUAL FLOWRATE}}{\text{NOMINAL FLOWRATE}}$$

$$= \quad \frac{\text{NOMINAL - SLIP}}{\text{NOMINAL}}$$

Hence:

$$\text{SLIP} \quad = \quad (1 - \eta_V) \times \text{NOMINAL}$$

$$\text{NOMINAL} \quad = \quad \frac{3000 \text{ rpm} \times 20.3 \text{ in.}^3/\text{rev}}{60 \text{ s/min}} = 101.5 \text{ in.}^3/\text{s}$$

$$\text{SLIP} \quad = \quad (1 - 0.965)101.5 = 3.553 \text{ in.}^3/\text{s}$$

We now need to know the oil viscosity that was used. This was not given for the chosen pump, but it is quite common to reduce all data to a standard viscosity of 100 SUS. The full expression for conversion of SUS to kinematic viscosity is given in ASTM D2161 [3] requires a root finding procedure. When this is done for SUS = 100, the result is:

$$\nu = 21.52 \, \frac{\text{mm}^2}{\text{s}}$$

A simpler expression obtained from Merritt [2] may be used without much loss of accuracy:

$$\nu \quad = \quad 0.216 \text{ SUS} - \frac{166}{\text{SUS}}$$

$$= \quad 0.216 \times 100 - \frac{166}{100} = 19.94 \, \frac{\text{mm}^2}{\text{s}}$$

Using the conversion procedure presented in Chapter 2 that with $\nu = 199.94$ mm/s and a specific gravity of 0.85 then the absolute viscosity in inch-pound force-second units is:

$$\mu = 2.45\text{E}{-}6 \ \frac{\text{lb}_\text{f}\cdot\text{s}}{\text{in.}}$$

Noting:

$$C_s = \frac{\mu}{D_m}(C_{im} + C_{em})$$

Where:

$$
\begin{aligned}
(C_{im} + C_{em}) &= \frac{Q_s}{p_1} = \frac{3.553 \text{ in.}^3/\text{s}}{3000 \text{ lb}_\text{f}/\text{in.}^2} \\
&= 1.184\text{E}{-}3 \ \frac{\text{in.}^5}{\text{lb}_\text{f}\cdot\text{s}}
\end{aligned}
$$

The specification sheet value of D_m must be converted to in.^{3}/rad:

$$D_m = \frac{2.03 \text{ in.}^3/\text{rev}}{2\pi \text{ rad/rev}} = 0.323 \ \frac{\text{in.}^3}{\text{rad}}$$

$$
\begin{aligned}
C_s &= \frac{2.45\text{E}{-}6 \text{ lb}_\text{f}\cdot\text{s/in.}^2}{0.323 \text{ in.}^3/\text{rad}}1.184\text{E}{-}3 \text{ in.}^5/\text{lb}_\text{f}\cdot\text{s} \\
&= 8.98\text{E}{-}9 \quad \text{(dimensionless)}
\end{aligned}
$$

8.4   FORM OF CHARACTERISTIC CURVES

8.4.1   Volumetric Efficiency

We shall first consider the form of volumetric efficiency vs. viscosity·speed/pressure (dimensionless speed). In practice, manufacturers will present volumetric efficiency vs. speed only and we shall show how this simplification is achieved. It was shown earlier that a real motor would have an inflow represented by:

$$Q_1 = D_m\omega + C_{im}(p_1 - p_2) - C_{em}(p_1 - p_0) \tag{8.7}$$

As indicated, manufacturers need a baseline condition and this is chosen as the condition where $p_1 = p_L$ and $p_2 = 0$. In this situation, the flow into a motor with leakage may be expressed as:

$$Q_1 = D_m \omega + C_s \frac{D_m}{\mu} p_1 \tag{8.14}$$

The volumetric efficiency for a motor is defined as:

$$\eta_{mV} = \frac{\text{NOMINAL FLOW}}{\text{ACTUAL FLOW}} = \frac{D_m \omega_m}{D_m \omega_m + C_s \frac{D_m}{\mu} p_1}$$

$$= \frac{1}{1 + C_s / \left(\frac{\mu \omega_m}{p_1} \right)} \tag{8.15}$$

First examine $\mu \omega_m / p_1$. Because C_s was shown to be dimensionless, then $\mu \omega_m / p_1$ should also be dimensionless:

$$\frac{\mu \omega_m}{p_1} = \frac{M}{LT} \frac{1}{T} \frac{LT^2}{M} = M^0 L^0 T^0$$

dimensionless as predicted. With some algebraic manipulation, it may be shown:

$$(\eta_{mV} - 1) = -\frac{C_s}{\left(\frac{\mu \omega_m}{p_1} \right) - (-C_s)} \tag{8.16}$$

If the common substitution x for the abscissa variable and y for the ordinate variable is made, then the equation of volumetric efficiency vs. dimensionless speed is:

$$(y - b) = \frac{-C_s}{(x - a)} \tag{8.17}$$

This function may be recognized as a rectangular hyperbola with the origin translated to (a, b) and mirrored in the x axis because of the minus sign in the numerator. Only the branch of the rectangular hyperbola above the x axis is relevant because η_V is necessarily positive. A generic volumetric efficiency curve is shown in Figure 8.10.

Although the analysis just performed is quite useful and certainly produces curves that show the same trends as those found on manufacturers' specification sheets, it will be found that data taken at several different pressures cannot be forced to lie on one curve as would be suggested by the analysis. This situation is not surprising. If we were to consider the leakage of oil through the gap between the slipper and the swash plate in an axial piston pump, then it seems likely that this leakage would be a function of pressure and speed.

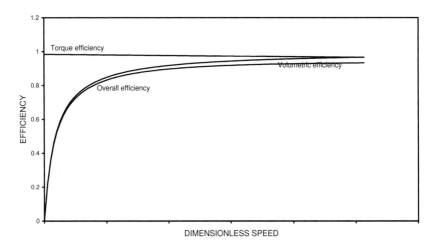

Figure 8.10: Volumetric, torque, and overall efficiencies vs. dimensionless speed $\mu\omega_m/p_1$.

Manufacturers use speed in revolution per minute as the abscissa rather than dimensionless speed. Because the volumetric efficiency will be affected by pressure and viscosity, the curve is presented at one standard viscosity. In the U.S. this would often be at 100 SUS. The effect of pressure is accounted for by presenting a family of curves at various pressures. Figure 8.11 shows the curves in the forms that might be presented by manufacturers. Here the abscissa values have been changed from dimensionless speed to speed in revolution per minute, the pressure is displayed as 1000, 2000, and 3000 $lb_f/in.^2$ and the oil viscosity is displayed as 100 SUS. As expected, increasing pressure will decrease volumetric efficiency because there will be more leakage flow. On the other hand, increasing speed will increase the nominal flow term and leave the leakage terms (at least in the simple analysis) unchanged so volumetric efficiency will increase as speed increases.

8.4.2   Torque Efficiency

There are three resisting torques on the shaft of a pump or motor, viscous resistance, sliding friction resistance, and shaft sealing resistance.

Viscous resistance: The torque from viscous drag is a function of viscosity and angular velocity. As with the analysis for volumetric efficiency,

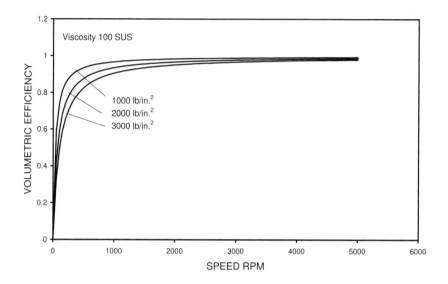

Figure 8.11: Manufacturers' method of presenting volumetric efficiency.

we shall not attempt to relate viscous drag to specific geometry. We shall consider a generic expression:

$$T_d = C_d \mu \omega_m \tag{8.18}$$

Checking the dimensions of Cd:

$$
\begin{aligned}
C_d &= \frac{T_d}{D_m \mu \omega_m} = \frac{\left(\frac{\mathrm{ML}^2}{\mathrm{T}^2}\right)}{\left(\mathrm{L}^3 \frac{\mathrm{M}}{\mathrm{LT}} \frac{1}{\mathrm{T}}\right)} \\
&= \mathrm{M}^0 \mathrm{L}^0 \mathrm{T}^0 \qquad \text{dimensionless}
\end{aligned}
$$

Sliding friction resistance: Although a pump or motor is well lubricated, there are still metal-to-metal contacts that contribute to Coulomb friction. It is generally assumed that the normal force between such rubbing parts will be directly proportional to pressure. Thus the expression for the drag torque due to Coulomb friction is:

$$T_f = \frac{\omega_m}{|\omega_m|} C_d D_m (p_1 + p_2) \tag{8.19}$$

The expression $\omega_m/|\omega_m|$ is a means of changing the sign of the friction torque to match the direction of rotation. If we examine:

$$C_f = \frac{T_f}{pD_m} = \frac{\frac{ML^2}{T^2}}{\frac{M}{LT^2}L^3}$$
$$= M^0L^0T^0 \qquad \text{dimensionless}$$

Shaft sealing resistance: The shaft sealing torque is regarded as independent of pressure and velocity (it must, however, change sign as the direction of rotation changes).

$$T_c = \frac{\omega_m}{|\omega_m|}T_c \qquad (8.20)$$

Note that the shaft sealing torque is usually small compared with the other two drag torques so it is often dropped from analyses.

Total expression for drag torque: The total drag torque on a pump or motor can be approximated by:

$$T_R = C_d D_m \mu \omega_m + C_f \frac{\omega_m}{|\omega_m|} D_m(p_1 + p_2) + \frac{\omega_m}{|\omega_m|}T_c \qquad (8.21)$$

Hence using the usual baseline criterion that manufacturer's specifications are given for $p_2 = 0$, then the torque delivered by a motor will be:

$$T_L = D_m p_1 - \left(C_d D_m \mu \omega_m + C_f \frac{\omega_m}{|\omega_m|} D_m p_1 + \frac{\omega_m}{|\omega_m|}T_c\right) \qquad (8.22)$$

For a motor rotating in one direction, we can write the torque efficiency as:

$$\eta_{Tm} = \frac{T_L}{D_m p_1}$$
$$= \frac{D_m p_1 - (C_d D_m \mu \omega_m + C_f D_m p_1 + T_c)}{D_m p_1} \qquad (8.23)$$

As indicated previously, the shaft sealing torque is usually sufficiently small that it can be ignored so:

$$\eta_{Tm} = (1 - C_f) - C_d \left(\frac{\mu \omega_m}{p_1}\right) \qquad (8.24)$$

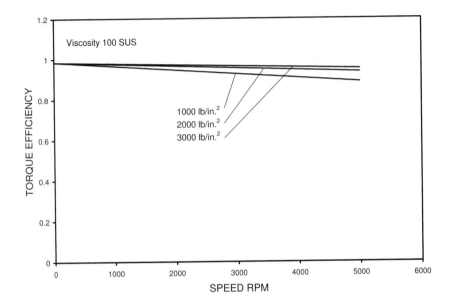

Figure 8.12: Manufacturers' method of presenting torque efficiency.

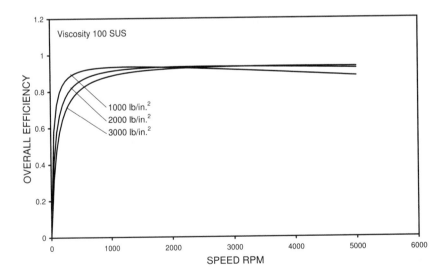

Figure 8.13: Manufacturers' method of presenting overall efficiency.

A graph of torque efficiency vs. speed is shown in Figure 8.12. The form of this graph is not immediately obvious. Intuitively, we might expect the efficiency to drop as the load pressure increases because the Coulomb friction torque will increase with pressure. The reason for the trends shown is that the absolute power increases as the pressure increases. Likewise the absolute viscous torque is independent of the pressure. Thus the fraction of the absolute torque dissipated as viscous drag torque decreases as the load pressure increases. That is the trend shown in the figure.

8.4.3 Example: Estimating Motor Performance

Given manufacturer's information on the torque efficiency of a specific motor and the information in Table 8.1, estimate the magnitude of C_d and C_f:

Table 8.1: Information for motor performance example

Characteristic	Value	Units
Motor displacement, D_m	0.323	in.^{3}/rad
Speed, n	2400	rpm
Torque efficiency η_T	96.9	%
Pressure p_L	3000	lb$_f$/in.^{2}
Oil absolute viscosity, μ	2.4E$-$6	lb$_f \cdot$ s/in.^{2}
Partition between viscous and Coulomb friction	0.5:0.5	

If the motor were perfect, the torque delivered would be:

$$p_L D_m = 3000 \, \frac{\text{lb}_f}{\text{in.}^2} \times 0.323 \frac{\text{in.}^3}{\text{rad}}$$
$$= 969 \text{ in.} \cdot \text{lb}_f$$

Hence the drag torque would be:

$$T_d = p_L D_m (1 - \eta_{Tm})$$
$$= 3000 \, \frac{\text{lb}_f}{\text{in.}^2} \times 0.323 \, \frac{\text{in.}^3}{\text{rad}} \times (1 - 0.969)$$
$$= 30.04 \text{ in.} \cdot \text{lb}_f$$

Thus:

$$C_d = \frac{0.5 \times 30.04 \text{ lb}_f \cdot \text{in.}}{0.323 \text{ in.}^3/\text{rad} \times 2.45\text{E}{-}6 \text{ lb}_f \cdot \text{s/in.}^2 \times 2\pi \text{ rad/rev} \times \frac{2400 \text{ rev/min}}{60 \text{ s/min}}}$$

$$= 75520$$

and:

$$C_f = \frac{T_f}{P_L D_m} = \frac{0.5 \times 30.04 \text{ lb}_f.\text{in.}}{3000 \text{ lb}_f/\text{in.}^2 \times 0.323 \text{ in.}^3/\text{rad}}$$

$$= 0.0155$$

8.4.4 Overall Efficiency

Overall efficiency plotted vs. speed at three load pressures is shown in Figure 8.13. At high speeds, the volumetric efficiency increases and the torque efficiency drops (Figure 8.10), so it is to be expected that there will be a speed at which overall efficiency shows a maximum. It may be observed that the efficiency ranking with pressure is that lowest pressure leads to highest efficiency at low speed, this ranking is reversed at high speed. This result could have been predicted by examining the volumetric and torque efficiency curves.

$$
\begin{aligned}
\eta_{OA} &= \frac{\text{POWER OUT}}{\text{POWER IN}} \\
&= \frac{T_L \omega_m}{p_1 Q_1} = \frac{T_m \omega_m}{p_1 \left(\frac{D_m \omega_m}{\eta_{Vm}} \right)} \\
&= \frac{T_L}{p_1 D_m} \eta_{Vm} = \eta_{Tm} \eta_{Vm}
\end{aligned}
\qquad (8.25)
$$

If the theoretical curves presented here are compared to curves for real equipment, the trends are very similar. One feature that should be observed is that fluid power pumps and motors show quite dramatic drops in efficiency if they are operated at slow speeds. The theoretical model and real equipment agree on this trend. For a more recent discussion of pump efficiency, the reader is referred to [4].

PROBLEMS

8.1 A fluid power pump is used to drive a motor as shown in the figure. Pressure loss, Δp_V, across the valve at position 3 is 12 $lb_f/in.^2$.

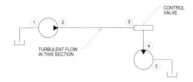

Determine the Reynolds number, Re, in the flow line between stations 2 and 3. Determine pressures, p_2 and p_3, for the conditions given in the table.

Characteristics of a pump and motor system

Characteristic	Size	Units
Oil mass density, ρ	0.08E−3	$lb_f \cdot s^2/in.^4$
Oil viscosity, μ	1.7E−6	$lb_f \cdot s/in.^2$
Line length, 2 to 4, ℓ_L	63.0	in.
Line diameter, 2 to 4, d	0.4	in.
Pressure at 1, p_1	0.0	$lb_f/in.^2$
Pump speed, n	1075	rpm
Pump overall efficiency, η_{op}	92.0	%
Pump mechanical efficiency, η_{mp}	95.0	%
Valve loss factor, K	5.1	
Motor displacement, D_m	1.95	$in.^3/rev$
Motor torque output, T_m	380	$lb_f \cdot in.$
Motor discharge pressure, p_5	27.0	$lb_f/in.^2$
Motor mechanical efficiency, η_{mm}	97.0	%

8.2 A fluid power pump is used to drive a motor as shown in the figure. Pressure loss, Δp_V, across the valve at position 3 is 14 $lb_f/in.^2$

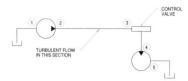

Determine the required speed, n_p for the pump. Determine the Reynolds number, Re, in the flow line between stations 2 and 3. Determine pressures, p_1, p_3, and p_4, for the conditions given in the table.

Characteristics of a pump and motor system

Characteristic	Size	Units
Oil mass density, ρ	0.08E$-$3	$lb_f \cdot s^2/in.^4$
Oil viscosity, μ	1.7E$-$6	$lb_f \cdot s/in.^2$
Line length, 2 to 4, ℓ_L	65.0	in.
Line diameter, 2 to 4, d	0.4	in.
Pressure at 1, p_1	50.0	$lb_f/in.^2$
Pump overall efficiency, η_{op}	92.0	%
Pump volumetric efficiency, η_{vp}	95.0	%
Pump input torque, T_p	390	$lb_f \cdot in.$
Pump displacement, D_p	1.75	$in.^3/rev$
Valve loss factor, K	5.2	
Motor discharge pressure, p_5	27.0	$lb_f/in.^2$
Motor mechanical efficiency, η_{mm}	97.0	%

8.3 The two rear drive sprockets on a crawler tractor are powered with hydraulic motors through a reduction gear set. The motors are driven by a single variable displacement hydraulic pump with equal flow to each motor.

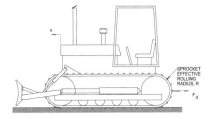

Determine the highest and lowest sprocket speeds, n_{sl} and n_{sh} and the lowest and highest tractor speeds, $\dot{x}$ and $\dot{X}$ that will occur as the pump displacement advances from D_{pl} to D_{ph}. Determine the tractor drawbar pull, F_d, and the tractor power, P, that will be produced with the values for pump displacement D_{pl} and D_{ph}, and the other values given in the table.

Characteristics of a hydrostatic transmission for a crawler tractor

Characteristic	Size	Units
Outlet pressure, p_s	23.5	MPa
Pump displacement, low, D_{pl}	25.0	mL/rev
Pump displacement, high, D_{ph}	55.0	mL/rev
Pump speed, n_p	1200	rpm
Pump volumetric efficiency, η_{vp}	97.0	%
Motor displacement, D_m	330	mL/rev
Motor volumetric efficiency, η_{vm}	97.0	%
Motor mechanical efficiency, η_{mm}	96.0	%
Motor discharge pressure, p_r	240	kPa
Gear ratio, $N = n_m/n_s$	4.5:1	
Sprocket effective rolling radius, r	375	mm

8.4 The two rear drive sprockets on a crawler tractor are powered with hydraulic motors through a reduction gear set. The motors are driven by a single variable displacement hydraulic pump with equal flow to each motor.

Determine the required motor displacement, D_m, to produce the given dozer force, F_z, and the drawbar force, F_d. Determine the required motor flow, Q, to produce the given tractor speed, $\dot{x}$.

Characteristics of a hydrostatic transmission for a crawler tractor

Characteristic	Size	Units
Drawbar force, F_d	12000	N
Dozer force, F_z	19500	N
Motor inlet pressure, p_s	30.0	MPa
Motor discharge pressure, p_2	350	kPa
Motor overall efficiency, η_{om}	94.0	%
Motor mechanical efficiency, η_{mm}	96.0	%
Gear ratio, $N = n_m/n_s$	3.9:1	
Tractor speed, $\dot{x}$	4.7	km/h
Sprocket effective rolling radius, r	377	mm

8.5 The two rear drive sprockets on a crawler tractor are powered with hydraulic motors through a reduction gear set. The motors are driven by a single variable displacement hydraulic pump with equal flow to each motor.

Determine the lowest and highest sprocket speeds, n_{sl} and n_{sh}, and the lowest and highest tractor speeds, $\dot{x}$ and $\dot{X}$, that will occur as the motor displacement changes from, D_{mh} to D_{ml}. Determine the tractor drawbar pull, F_d, and the tractor power, P, that will be produced with the values for motor displacement, D_{ml} and D_{mh}, for the values given in the table.

Characteristics of a hydrostatic transmission for a crawler tractor

Characteristic	Size	Units
Pump outlet pressure, p_s	23.5	MPa
Pump displacement, D_p	40.0	mL/rev
Pump speed, n_p	1200	rpm
Pump volumetric efficiency, η_{vp}	97.0	%
Motor low displacement, D_{ml}	220	mL/rev
Motor high displacement, D_{mh}	570	mL/rev
Motor volumetric efficiency, η_{vm}	97.0	%
Motor mechanical efficiency, η_{mm}	96.0	%
Motor discharge pressure, p_r	240	kPa
Gear ratio, $N = n_m/n_s$	4.5:1	
Sprocket effective rolling radius, r	375	mm

8.6 A fluid power pump is used to drive a motor as shown in the figure. Pressure loss, Δp_V, across the valve at position 3 is 14 lb$_f$/in.^{2}

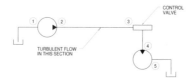

Determine the required displacement, D_p for the pump. Determine the Reynolds number, Re, in the flow line between stations 2 and 3. Determine pressures, p_2 and p_3, for the conditions given in the table.

Characteristics of a pump and motor system

Characteristic	Size	Units
Oil mass density, ρ	0.08E−3	$lb_f \cdot s^2/in.^4$
Oil viscosity, μ	1.7E−6	$lb_f \cdot s/in.^2$
Line length, ℓ_L	65.0	in.
Line diameter, d	0.375	in.
Pressure at 1, p_1	0.0	$lb_f/in.^2$
Pump speed, n	1150	rpm
Pump overall efficiency, η_{op}	94.0	%
Pump mechanical efficiency, η_{mp}	95.0	%
Valve loss factor, K	5.7	
Motor displacement, D_m	2.1	$in.^3/rev$
Motor torque output, T_m	385	$lb_f \cdot in.$
Motor discharge pressure, p_5	33.0	$lb_f/in.^2$
Motor mechanical efficiency, η_{mm}	95.0	%

REFERENCES

1. Karassik, I. J., Krutzsch, W. J., Fraser, W. H., and Messina, J. P., 1976, *Pump Handbook*, McGraw-Hill Book Company, New York, NY.

2. Merritt, H. E., 1967, *Hydraulic Control Systems*, John Wiley & Sons, New York, NY.

3. ASTM International, 1998, "Standard Practice for Conversion of Kinematic Viscosity to Saybolt Universal Viscosity or to Saybolt Furol Viscosity", D2161-93, West Conshohocken, PA.

4. Manring, N. D., 2005, *Hydraulic Control Systems*, John Wiley & Sons, Inc., Hoboken, NJ.

9

AXIAL PISTON PUMPS AND MOTORS

9.1 PRESSURE DURING A TRANSITION

In Chapter 8, it was mentioned that the design of the pressure transition grooves in the valve plate of an axial piston pump or motor determines the form of the pressure vs. angle curves as a cylinder passes through top or bottom dead center. Although a systems engineer will probably never need to design the pressure transition groove geometry, some knowledge of its function and geometry may be useful.

In an axial piston machine, the port at the end of the pumping cylinder passes over a series of kidney shaped ports in the valve plate. At positions away from the top and bottom dead center positions, the solid portions between the valve plate ports are significantly smaller than the width of the cylinder. Thus there is negligible flow attenuation during a transition. At the top and bottom dead center positions, however, the situation changes. Overlap here would lead to connection between the low pressure suction port and the high pressure discharge port. Ideally, the pressure in the cylinder should change instantly from one extreme to the other at the top and bottom dead center positions. Such a pressure transition is not achievable in practice because unrealistic manufacturing tolerances would be required. In practice, manufacturers provide controlled leakage paths by machining small tapered grooves in the valve plate. The cylinder is allowed to communicate simultaneously with the high and low pressure ports, but the connecting passages are limited in size.

The analytical method used to determine pressure vs. angle is generally

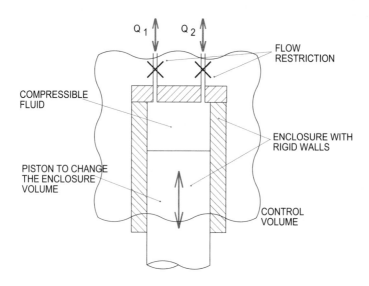

Figure 9.1: Control volume for analyzing pressure change in cylinder.

applicable to many fluid power components. In this analysis, we shall
consider the pressure transition in a pump. A motor would be analyzed
in a similar fashion except the flows from the high and low pressure lines
would be reversed and the piston would be moving from the low to high
side at top dead center.

Consider the volume of oil trapped in the cylinder bore as the piston
reaches the end of its inward stroke. A control volume shown in Figure 9.1.
The volume of the fluid trapped in the cylinder is changing because the
piston is moving. There are potentially two ports through which fluid can
pass in or out of the cylinder. The first port is the connection to the high
pressure discharge line and the second port is the connection to the low
pressure suction line. For this analysis it will be assumed that the discharge and suction pressures are invariant. As indicated in the discussion
of the axial piston pump in Chapter 8, triangular form pressure transition
grooves are present in the valve plate at top and bottom dead center allowing simultaneous passage of fluid to both the discharge and suction lines
for some small angular rotation of the piston barrel. The last part of the
development, and perhaps the most important, is the fact that liquids are
slightly compressible. That is they exhibit bulk modulus (Chapter 2). To
achieve conservation of volume, we can write:

Change in cylinder volume =

Internal accumulation due to bulk modulus effects

+Total flow out of the cylinder (9.1)

The extension of the definition of bulk modulus to a dynamic expression for compliance was manipulated in Chapter 4 and the result is restated here:

$$\frac{dp}{dt} = \frac{\beta}{V}\frac{dV}{dt} \qquad (4.1)$$

Thus the differential equation accounting for compliance for this particular situation is:

$$\frac{dp}{dt} = \left(\frac{\beta}{V}\right)(Av(t) - Q_1(t) - Q_2(t)) \qquad (9.2)$$

In the current situation, the velocity of the piston at any time is treated as fully determined because the pump is driven at constant speed. We must make a suitable approximation for the conditions governing $Q_1(t)$ and $Q_2(t)$. An exact analysis of transition groove flow suitable for a pump manufacturer is not our goal. In practice, the flow in and out of the cylinder will be through a passage of finite length with a fairly complex geometry. A standard approach commonly employed in the fluid power field is to assume that flows through small constrictions can be treated as flows through circular, sharp edged orifices obeying the relationship:

$$Q = C_d A \sqrt{\frac{2\Delta p}{\rho}} \qquad (3.19)$$

In a commercial pump or motor, the main ports connecting the cylinder to the inlet and outlet lines would not be circular but kidney shaped (Figure 8.6). For the current analysis, the ports will be taken as circular. As a cylinder nears top dead center, the discharge is fleetingly a complete circle with a triangular appendage (the pressure transition groove). The full circle changes to two arcs of a circle and the triangular appendage. Finally flow is through a triangular port that is continuously diminishing in size. This situation is mirrored by the suction port, but the areas become larger as the cylinder progresses. A sequence is shown in Figure 9.2.

9.1.1   Simulation of the Pressure Transition

The various methods of performing fluid power simulations have been discussed in Chapter 4. This problem can be approached by the formulation

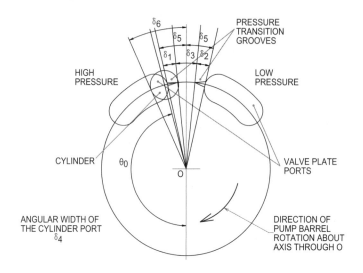

Figure 9.2: Valve plate and cylinder port geometry.

of the basic equations (Equations 9.2 and 3.19). The equations were formulated and solved using Visual Basic for Applications® program within
Excel®. As indicated in Section 9.1, analyzing the dynamic performance of
fluid power systems involves developing and solving sets of ordinary differential equations. These equations are often *stiff* [1] and may need special
techniques for integration. This is where the specialized simulation programs have an edge over using a simple user written routine. Developing
the equations is easier and solvers are available to handle the stiffness. On
the other hand, the specialized programs may conceal some analytical steps
from the user. The Runge-Kutta 4th order method of solving differential
equations was able to integrate this problem.

That there may be problems solving fluid power differential equations is
easily seen by examining Equation 9.2 again. In the portions of the analysis
where the ports have a relatively large area, the pressure drop across the
port will be low. This will be true from one simulation interval to the
next, consequently, the rate of pressure change, dp/dt will be small. This
small quantity is calculated from the product of a large quantity, β/V,
multiplied by a small quantity $Av(t) - Q_1(t) - Q_2(t)$. This last quantity
is the difference between the rate of change of the cylinder volume and the
flows out of the cylinder. As indicated earlier, this difference will become
vanishingly small when the port size is large. That is away from top or

bottom dead center. Difficulties might be expected with any integration routine under these conditions. The program overcomes this problem by breaking the simulation into three parts. Plain orifice flow without any differential equation is used at the beginning and end of the simulation where the port areas are large. The differential equation approach is only used during the transition where dp/dt will be quite large.

9.1.2 Results of the Simulation

The pump or motor designer must compromise. If there is no simultaneous connection of the high pressure and low pressure lines to a cylinder at top or bottom dead center, then high volumetric efficiency will be retained. On the other hand, such an approach will lead to a small volume of fluid being trapped in the cylinder during compression. Although fluids exhibit bulk modulus, the bulk moduli are high enough that undesirable high pressures will be developed. Such pressures may damage the device and may lead to a reduction in torque efficiency.

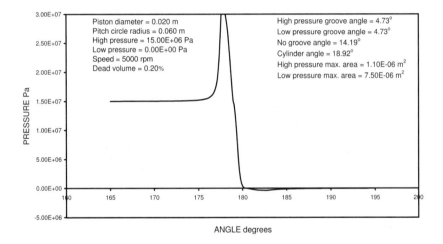

Figure 9.3: Example of incorrectly designed pressure transition grooves.

Figure 9.3 shows the pressure history during the transition of a cylinder in a pump that would be considered unsatisfactory. The pressure rises to double the nominal working pressure before top dead center and falls to nearly zero absolute afterwards. The low pressure at the beginning of the suction stroke could lead to cavitation. The simulation was performed for a pump with 20 mm diameter pistons and a pitch circle radius of 60 mm. Ob-

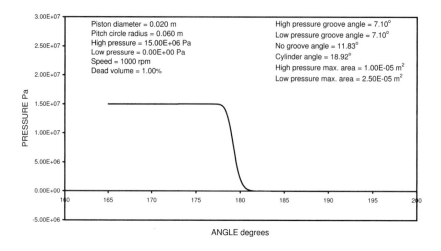

Figure 9.4: Acceptable low speed transition for axial piston pump.

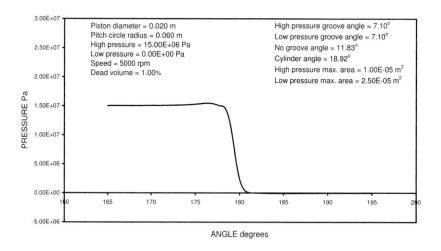

Figure 9.5: Acceptable high speed transition for axial piston pump.

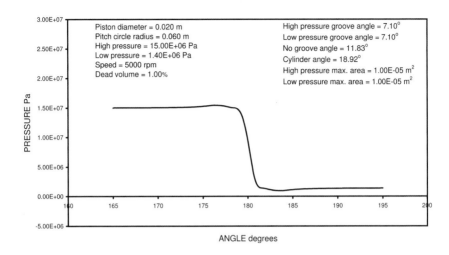

Figure 9.6: Acceptable high speed transition for a closed circuit axial piston pump.

serve that the pressure transition grooves only extend 25% into the cylinder diameter when fully open. Also the angular distance between the tips of the discharge and suction pressure transition grooves is 75% of the cylinder diameter. The chosen dimensions do not permit trapping of fluid with no discharge, but the passages available are too small to limit the pressure rise adequately. The areas of the transition grooves were deliberately made small. The high pressure side would be a triangle 0.22 mm at the base and 5 mm along the cylinder pitch circle. The low pressure side would be 1.5 mm at the base and 5 mm along the cylinder pitch circle.

Another feature of pumps and motors may now be addressed. In hydrostatic transmissions, the suction side may be maintained at a pressure significantly above atmospheric pressure (closed circuit operation). For the simulation example the low pressure was deliberately set at zero gauge (open circuit operation). If cavitation is to be avoided, then the pressure in the cylinder should be above absolute zero at all times. For this example, the pressure transition groove for the low pressure side is much larger than that on the high pressure side. Inspection of the valve plate of an axial piston pump or motor will show if the device was designed for open or closed circuit operation.

Changes were made to several characteristics to achieve satisfactory operation at both high and low speeds. The pressure transitions curves for

1000 rpm and 5000 rpm for a well designed pump are shown in Figures 9.4 and 9.5. As indicated initially, a compromise must be made. A small pressure rise at high speed is the penalty that must be paid to avoid excessive leakage from high to low side at low speed. Figure 9.6 shows the pressure transition that would be obtained for a closed circuit pump in which the high and low pressure transition grooves have been set to have equal areas. Keeping the areas as small as possible will help maintain volumetric efficiency, but the penalty is a moderate undershoot of pressure. Because this is a closed circuit pump with a charge pump keeping the suction side pressure above atmospheric pressure, there is no danger of cavitation so the pressure undershoot is acceptable.

One more comment on the analysis. There is a residual or dead volume V_0 representing the cylinder volume at top or bottom dead center that is introduced into the simulation. This volume can be very small, but *must not be zero* or else the simulation will fail. Essentially there will be division by zero and the rate of pressure change would be infinite.

9.2   TORQUE AFFECTED BY PRESSURE TRANSITION – AXIAL PISTON PUMP

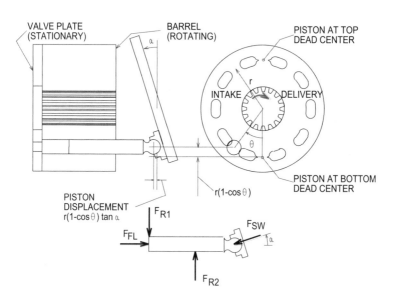

Figure 9.7: Nomenclature for calculating the torque on the barrel of an axial piston pump.

A free body diagram showing the forces on the piston of a pump or motor is shown in Figure 9.7. Note that sliding (i.e. Coulomb) friction between the piston and the bore is being ignored. We shall use this diagram to perform two calculations. In the first we shall calculate the torque on the barrel of an ideal, frictionless pump by analysis of the forces on the barrel as it rotates. This will be a useful introduction to the second calculation where a non ideal pressure transition will be selected and we shall use a computer program to show that the torque needed by the pump differs from the ideal quantity. Consider the forces on the piston shown in Figure 9.7. We are analyzing an ideal, frictionless machine so there will be no friction induced forces axially on the piston due to the two reaction forces F_{R1} and F_{R2}, also only a normal force will be considered at the junction of the swash plate and the slipper (F_{SW}) for the same reason. The force caused by the fluid pressure acting on the piston face will be:

$$F_{FL} = pA$$

For this analysis, we are only interested in the force acting on the barrel that must be overcome to cause the pump to operate. For the ideal pump without friction, this force always acts in a plane parallel to the plane containing the shaft axis and the top and bottom dead centers. Resolving forces axially yields:

$$F_{SW} \cos \alpha = F_{FL}$$
$$F_{SW} = \frac{F_{FL}}{\cos \alpha}$$

The force on the barrel that must be overcome by the shaft drive torque is:

$$F_{R2} - F_{R1} = F_{SW} \sin \alpha$$
$$= \frac{F_{FL}}{\cos \alpha} \sin \alpha = F_{FL} \tan \alpha$$

The instantaneous torque on the barrel at some angle θ from bottom dead center is:

$$T_\theta = (F_{R2} - F_{R1})r \sin \theta = F_{FL}r \tan \alpha \sin \theta$$
$$= (p_l A r \tan \alpha) \sin \theta \qquad (9.3)$$

Now keep the calculation basic and assume that $p_1 = p_L$ and $p_2 = 0$. The average torque that must be applied to the pump shaft (i.e., to the barrel) will be:

$$T_g = \frac{\int_0^\pi Ar \tan\alpha \sin\theta d\theta}{2\pi}$$

$$= p_L \frac{Ar \tan\alpha}{\pi} \qquad (9.4)$$

The pump displacement per revolution (examine Figure 9.7) is:

$$= 2Ar \tan\alpha$$

Thus the pump displacement per radian is:

$$D_p = \frac{2Ar \tan\alpha}{2\pi} = \frac{Ar \tan\alpha}{\pi}$$

Consequently the average torque required to operate the pump is the familiar form (Equation 8.4) with the D subscript changed from m to p:

$$T_g = D_p p_L \qquad (8.4)$$

9.2.1   Effect on Torque if the Pressure Change at Transition Is not Immediate

As we discussed in Section 9.1, the geometry of the pressure transition grooves affects the pressure change profile at top and bottom dead center. Another Visual Basic for Applications® program was used to calculate the torque on the piston barrel for an arbitrary pressure profile. In essence, this requires a table of pressure vs. angle for 0° to 360°. The program reads this file into an array and performs linear interpolation to obtain pressure at any angle as the program calculates the torque throughout a complete revolution. Figure 9.8 shows an idealized pressure distribution where the pressure changes instantaneously at top and bottom dead center. The torque curve derived from the program is shown in Figure 9.9. The shape of this curve is that of a half sine curve. This is the result that would be expected from examining Equation 9.3.

The pressure distribution shown in Figure 9.3 (shown again over one complete revolution in Figure 9.10) was used to generate a pressure vs. angle table used to derive the torque vs. angle curve shown in Figure 9.11. Although the pressure spike is double the working pressure and would not be considered good pump design, the effect on the torque is minimal. The reason is that the excessive pressure only occurs at top and bottom dead center where the lever arm of the transverse forces on the piston are very small.

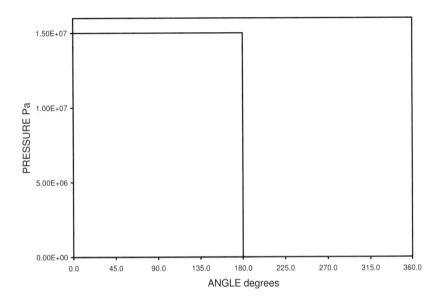

Figure 9.8: Idealized pressure distribution for an axial piston pump.

9.3 TORQUE AND FLOW VARIATION WITH ANGLE FOR MULTICYLINDER PUMPS

First consider the flow rate from a single cylinder pump. Figure 9.7 shows that the piston displacement from bottom dead center is:

$$x = r(1 - \cos\theta)\tan\alpha$$

The flow rate out of the pump is given by:

$$
\begin{aligned}
Q = AV &= (Ar\tan a)\frac{d}{dt}(1 - \cos\theta) \\
&= (A\dot{\theta}r\tan\alpha)\sin\theta \qquad (9.5)
\end{aligned}
$$

Observe that the terms in the parentheses are constant for a given pump size, swash plate angle, and speed. Thus the equation is identical in form to Equation 9.3. The form of this equation was presented in Figure 9.9.

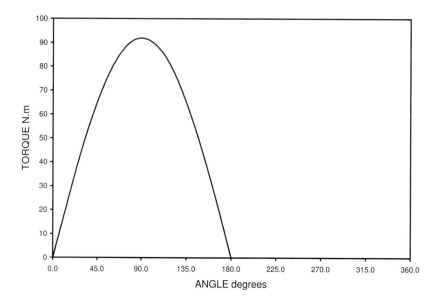

Figure 9.9: Torque required to drive the barrel of an ideal, single piston, axial piston pump.

Pumps and motors that only employed a single cylinder would operate quite roughly because of the asymmetry between the inlet and discharge portions. The results presented in Figure 9.12 have been normalized so the mean torque or flow rate for a single cylinder pump or motor has a value of one. As to be expected, adding more pistons increases the torque or flow rate absorbed or generated. A feature that should be noted is that the ripple pattern from even number and odd number pistons is different. The fundamental frequency of even number piston devices matches that of the number of pistons. On the other hand, the fundamental frequency of odd numbered piston devices is twice the number of pistons. Another important feature that may be observed from Figure 9.12 is that the amplitude of the torque or flow rate variation for odd numbered devices is much less than that for the even numbered devices. Consequently all manufacturers of axial piston devices only provide devices with odd numbers of pistons. Units that must be built to a price, are often only provided with five pistons. Better quality units will normally use nine or 11 pistons.

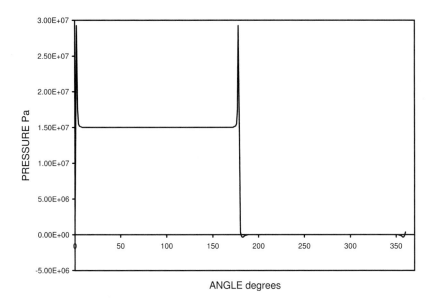

Figure 9.10: Example of unsatisfactory pressure distribution for an axial piston pump.

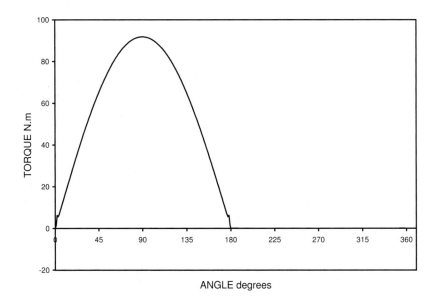

Figure 9.11: Torque required to drive a single piston, axial piston pump with unsatisfactory pressure distribution.

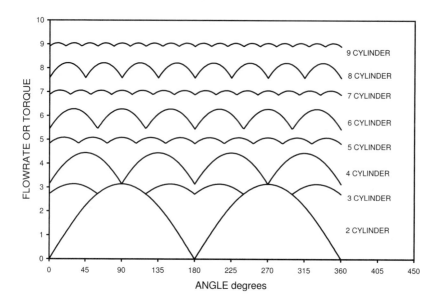

Figure 9.12: Torque or flow rate variation with angle for multicylinder axial piston devices.

9.3.1 Noise

Another reason for only manufacturing devices with an odd number of pistons is noise. Consider the equation for flow through a sharp edged orifice and rearrange this in the form:

$$p = \rho \left(p_o + \frac{1}{2} \left(\frac{Q}{C_d A} \right)^2 \right) \tag{9.6}$$

If p and Q are regarded as functions of time, then this expression could be used to predict the variation in pressure downstream of a pump forcing fluid through an orifice against a fixed downstream pressure of p_o. Consequently if Q varies in the manner shown in Figure 9.12, then p will vary in a similar fashion. If the line to the orifice is a flexible hose, then the walls of the hose will flex and radiate sound. Suppose there is a nine cylinder pump that is driven by a motor running at about 1800 rpm. The fundamental frequency of the sound generated by the pressure variation will be:

$$f = \frac{1800 \times 18}{60} = 540 \text{ Hz}$$

Because the ripple is not sinusoidal, it is likely that the first harmonic, 1080 Hz, will also be noticeable. It is generally accepted that hydraulic systems are noisy so any contribution to reducing flow variation will assist noise control.

PROBLEMS

9.1 A double acting cylinder is used to raise the mass, m. The hydraulic pump has seven equally spaced pistons.

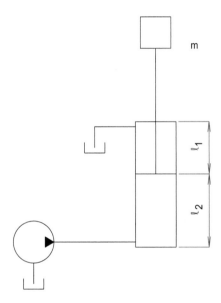

Determine the spring rates, k_1 and k_2 caused by the oil compressibility in cylinder lengths, ℓ_1 and ℓ_2. Determine the pump speed, n, at which the pump output pulse frequency, f_p, is equal to the natural frequency, f_n, of the cylinder and mass, m. Determine the pump speed, n, at which the pump output pulse frequency, f_p, is equal to the natural frequency, f_n, of the cylinder and mass, m, when the lengths, ℓ_1 and ℓ_2, are equal to 7 and 4 in. respectively.

Characteristics of double acting cylinder system

Characteristic	Size	Units
Piston diameter, d_p	3.5	in.
Rod diameter, d_r	1.1	in.
Length, ℓ_1	5.0	in.
Length, ℓ_2	6.0	in.
Mass, m	500	lb$_m$
Oil bulk modulus, β	175.0E+3	lb$_f$/in.^{2}

REFERENCES

1. Press, W. H., Flannery, B. P., Teukolsky, S. A., Vetterling, W. T., 1986, *Numerical Recipes The Art of Scientific Computing*, Cambridge University Press, Cambridge, U.K.

10

HYDROSTATIC
TRANSMISSIONS

10.1 INTRODUCTION

The function of any form of transmission is to match a power source, also called a prime mover, to a driven device. Transmissions serve several purposes, but the major purposes are to allow physical separation between the motor and the driven device, to scale torque, and to scale speed. Obviously, the transmission torque and speed cannot be scaled independently. Transmissions may also be characterized in terms of the steps of speed ratio between the motor and the driven device. The speed ratio for drives using sheaves or gears is usually fixed by geometry and will consist of one or more fixed ratios. In some applications, it is important that the speed ratio can be varied continuously over some desired range during operation. This can be achieved by a stepless transmission.

Consider a hoist. Functionality would be improved for any but the simplest hoist if the load can be lifted from rest slowly and without jerking. When in motion, the load should be accelerated smoothly. Consider first applying the form of three fixed gear ratios in the form of spur gears. Smooth transition between gear ratios is achieved by linking the motor and the gearbox with a progressive action clutch. As the hoist is started from rest, the operator would use each gear over a range of engine speed from idle to some accepted maximum speed. Once the engine had reached the accepted maximum speed, the clutch would be disengaged and the gearbox shifted to the next gear to allow the hoist to lift faster at a lower engine speed. Although the load could be accelerated smoothly from rest, every time the clutch was disengaged, torque transmission between the prime mover and the hoist cable drum would be interrupted and the load could begin to descend. In some power transmission applications, some deceleration during a gear change is of little consequence. Allowing the load to descend would not be acceptable for the hoist. Thus a variable speed hoist needs a variable speed transmission in which torque can always be transmitted between the motor and the load even when the speed ratio is being changed. This can be achieved with a stepless transmission.

There are many different variable speed, stepless transmissions. Some that may be commonly encountered are sheaves in which one or both sheaves can have the belt pitch circle diameter varied during operation. Coaxial disks with a moveable friction wheel in between can achieve similar stepless speed change. These latter two geometries are usually limited to low power transmissions.

As the power and torque that must be transmitted increase, two other forms of stepless transmission will be found. The prime mover can drive an electrical generator that can have its electrical output controlled. This variable output is used to power an electric motor. Such drives are used in a railroad locomotive. One advantage in this application is that one generator can drive several lower capacity motors attached to each axle of the locomotive. This configuration would maximize the traction that can be achieved by the unit.

Finally, we come to the hydrostatic transmission. Here the prime mover drives a variable displacement pump. In turn, the high pressure oil is passed to a motor. Motors with fixed displacement are most common, but variable displacement motors may be used for a hydrostatic transmission requiring a very wide speed ratio range.

The pump section of a hydrostatic transmission is an axial piston pump with either a tilting swash plate or bent axis for displacement control. Because hydrostatic transmissions may be found in devices from small riding lawnmowers to large earthmoving equipment, the means of controlling the swash plate angle can be direct mechanical or via an electrically operated pilot valve. Obviously cost is a factor. Adding a pilot valve adds to cost, but also gives the designer much more flexibility in terms of the link between the operator and the pump. Axial piston pumps are often paired with axial piston motors, but any type of motor can be used. Some motors may exhibit lubrication problems when operated at very low speeds, so extreme operating conditions should be discussed with the motor manufacturer.

Hydrostatic transmissions are very popular in off road and agricultural equipment because they provide a satisfactory combination of compactness, cost, and location flexibility. Like the electric generator/electric motor combination they allow the designer considerable flexibility in locating the prime mover with the pump coupled to it and the motor, which may be fairly distant.

10.2 PERFORMANCE ENVELOPE

A hydrostatic transmission may be limited by torque at low output speeds and by power at high output speeds [1]. The torque limit is a result of the establishment of the maximum pump working pressure that will be set by the pump manufacturer. The power limit is more obvious. The output power can only equal or be less than the input power. In practice, the output power will always be less than the input power because of losses.

First consider an ideal variable displacement pump having no friction and no leakage. This pump drives an ideal motor. Let the maximum pump pressure set by the manufacturer be p_{max}. As will be discussed later, a practical hydrostatic transmission will operate with a non zero pressure, the charge pressure, in the low pressure side. Since we are only examining an ideal condition, consider the charge pressure to be zero. Hydrostatic transmissions typically operate with the prime mover set to work at a fixed speed, usually the speed at which the power is a maximum for an internal combustion engine. Let this speed be ω_p. In general we can write the power delivered by the pump as:

$$P = \omega_p D_p p_L \qquad (10.1)$$

Thus if the pump displacement, D_p, is initially zero, the output power from the system will be zero. If the pump displacement is increased slightly, there will be flow through the system. We shall assume that the load on the motor is such that the pressure in the high pressure side is p_{max}. Using the two forms of power expression and remembering that the system is taken as ideal (lossless), then:

$$P = p_{max} D_m \dot{\theta}_m = T_m \dot{\theta}_m \qquad (10.2)$$

Combining Equation 10.1 and 10.2 yields:

$$\dot{\theta}_m = \frac{D_p}{D_m} \omega_p \qquad (10.3)$$

The motor Equation 10.2 yields:

$$T_m = p_{max} D_m \qquad (10.4)$$

Because p_{max} and D_m are constant, then the output torque is constant and at a maximum. There will be a motor speed given by:

$$\dot{\theta}_m = \frac{P}{p_{max}} D_m \qquad (10.5)$$

at which the output power equals the input power. Any further increase in
motor speed will be accompanied by a drop in torque, so the output power
remains constant. The relation between output torque and output speed
will be given by:

$$T_m = \frac{P}{\dot{\theta}_m} \tag{10.6}$$

This is the equation of a rectangular hyperbola. Finally the output speed
reaches the value:

$$\dot{\theta}_{mmax} = \frac{D_{pmax}}{D_m} \dot{\theta}_p \tag{10.7}$$

The output speed cannot exceed this value. The output envelope of a

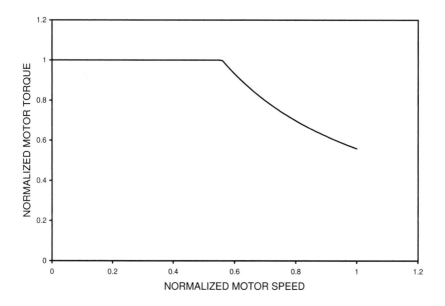

Figure 10.1: Output torque envelope for a hydrostatic transmission.

hydrostatic transmission is shown in Figure 10.1. It should be appreciated
that the output envelope shown in Figure 10.1 represents a system where
the power capability of the prime mover is less than the capability of the
hydrostatic transmission. If the power available from the prime mover
equals or exceeds the power capability of the pump given by Equation 10.1
then the output torque can remain constant up to the limiting speed of the
motor.

10.3 HYDROSTATIC TRANSMISSION PHYSICAL FEATURES

Although the mandatory components of a hydrostatic transmission are a variable displacement pump and a fixed displacement motor, other components are necessary to form a practical system. As with all fluid power systems, protection against excessive pressure surges should be provided. The potential for cavitation in the low pressure side should be addressed by maintaining this low pressure side significantly above atmospheric pressure. Most hydrostatic transmissions need to be able to work as well in reverse as forward. Another feature that is generally provided on a higher power system is cooling using a heat exchanger. All fluid power systems degrade some pressure energy into heat, but low power systems, e.g., a small riding lawn mower, might depend on air flow over an exposed reservoir for adequate cooling.

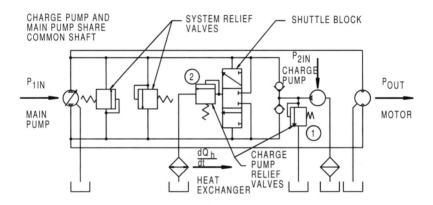

Figure 10.2: Basic components of a hydrostatic transmission.

These requirements alluded to in the previous paragraph are addressed by the various components shown in Figure 10.2. Two back to back relief valves are provided, so the current high pressure side can vent to the low pressure side. A relief valve that is exposed to a low pressure at its inlet and a high pressure at its outlet will block flow, so only one of the relief valves can function at a time.

Cavitation is prevented by providing a charge pump. This pump serves two purposes, first it maintains the low pressure side of the transmission at about 1.5 MPa and secondly it continually pumps oil through the system and through the heat exchanger. A variable displacement pump suitable for closed loop operation in a hydrostatic transmission will have small internal gear or gerotor pump built into the pump and mounted on an extension of the main pump shaft.

Note that the charge pump can pass oil into the low pressure side of the transmission through one of the two check valves. The charge pump oil leaves the loop through the shuttle block and pressure relief valve 2. Valve 2 is set at about 10% less pressure than valve 1 and operates continuously as long as the main pump is providing output.

Any instant when the load pressure in the transmission is reversed, then the connection of valve 2 must be switched from the old to the new low pressure side of the transmission through action of the shuttle block. The shuttle block spool is moved by the change in oil differential pressure caused by reversing the pump swash plate. There may be a period during which the shuttle block is in a neutral position and no oil can flow from the charge pump to the drain through valve 2. Because the charge pump is a positive displacement pump that is driven from the pump shaft, relief valve 1 serves to vent the charge pump flow to the drain under such conditions. The fact that this valve is set to a slightly higher pressure than valve 2 means that charge pump flow normally passes through valve 2 even though valve 1 is always exposed to the charge pump discharge.

Other features on Figure 10.2 are the case drains to the reservoir needed from both the pump and the motor. Also the charge pump suction is provided with a filter. Filters are seldom if ever provided in the transmission loop. First they would have to accommodate reversing flow and secondly they would have to operate at high pressure. The flow through the system provided by the charge pump serves to flush contaminants from the loop.

10.4 HYDROSTATIC TRANSMISSION DYNAMIC ANALYSIS

There may be occasions when the dynamic performance of a system must be evaluated. For example, if a hydrostatic transmission were being used in a system that positions a load, the designer might be concerned about potential overshoot occurring during the transition from a decelerating condition to a stationary condition.

We shall now analyze the specific example shown in Figure 10.3. A soil bin mounted on steel wheels running on a rail track is connected to a capstan mounted on the output shaft of a worm gearbox. The wire rope

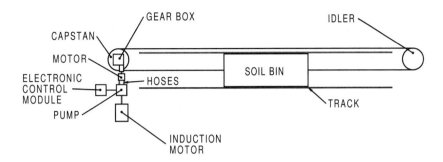

Figure 10.3: Application of a hydrostatic transmission to a research soil bin drive.

pulling the bin passes around an idler puller at the far end of the track, so the bin can be moved in both directions. The reverse motion, however, is only required to reposition the bin, and the combination of speed and accuracy is less critical for the return path. Only the forward motion will be analyzed here.

The system was designed to test tillage tools, so the bin was accelerated to a desired velocity, maintained at constant velocity while the tool was engaged, and then decelerated to rest. The worm gear box input was driven by the motor of a hydrostatic transmission. The prime mover for the system was an AC induction motor. The swash plate angle, i.e., the pump displacement, was controlled by an electrically operated pilot valve. The acceleration, constant velocity, and deceleration electrical inputs were generated by an electronic control box.

A full analysis of this system incorporating the dynamics of the valve and the pump is possible [2], but unnecessary in this context. The acceleration of the bin is much slower than the acceleration of the swash plate, so the analysis can be considerably simplified by ignoring the pump control dynamics.

Another feature that can be simplified is the torque and speed variation of the induction motor during operation. It may be shown [3], that an induction motor provides a driving torque when the shaft speed is slightly less than the synchronous speed of the supply and a resisting torque when the shaft speed exceeds synchronous speed. That is, an induction motor works as a generator when driven above the synchronous speed. The change in speed from the synchronous speed is only a few percent, so the induction motor is ideal for this application because it can handle acceleration and deceleration without external braking.

The analysis will only be performed for the basic condition where there is no tool engagement with the soil in the bin. Two further simplifications will be made to reduce the complexity of the analysis. Any backlash in the gearbox will be ignored and the elasticity of the driving cable will also be ignored. Because this is a fluid power device, leakage from the pump and motor will be considered and the effective bulk modulus of the oil in the hoses connecting the pump to the motor will be considered.

This problem is analyzed in a similar fashion to most of the fluid power problems analyzed in this text. The analysis is partitioned into one part that incorporates the pressure and rate of pressure change in the fluid portion of the system and a second part where the pressure acting on a motor element is expressed as torque. The acceleration caused by this torque acting on an inertia is then analyzed using Newton's Second Law, $F = ma$, or its angular equivalent. The relationship between the hydrostatic motor output angular velocity and the bin linear velocity must be determined:

$$\dot{\theta}_m = \frac{2N_{grbx}v_{bin}}{d_{cap}} \tag{10.8}$$

This may be written in a slightly more convenient form based on the linear displacement of the bin, x_{bin}:

$$\dot{\theta}_m = \left(\frac{2N_{grbx}}{d_{cap}}\right)\dot{x}_{bin} \tag{10.9}$$

Now consider the basic hydrostatic transmission. Fluid is supplied by an ideal pump at a rate of $\omega_p D_p$ where ω_p is considered fixed at the synchronous speed of the electrical supply when applied to the electric motor and D_p is a function of time. This input flow is partitioned into three: the leakage to the drain, the transient accommodation due to bulk modulus effects, and the flow through the motor:

$$\omega_p D_p = C_L p_L + \frac{V}{\beta_e}\frac{dp_L}{dt} + \dot{\theta}_m D_m \tag{10.10}$$

Write this equation as:

$$\frac{dp_L}{dt} = \frac{1}{V/\beta_e}\left(-C_L p_L - D_m\dot{\theta}_m + \omega_p D_p\right) \tag{10.11}$$

In this problem, we are not interested in the motor shaft angular velocity, $\dot{\theta}_m$. Consequently, we rewrite the equation using $\dot{x}$:

$$\frac{dp_L}{dt} = \frac{1}{V/\beta_e}\left(-C_L p_L - D_m\left(\frac{2N_{grbx}}{d_{cap}}\right)\dot{x} + \omega_p D_p\right) \tag{10.12}$$

Now we analyze the force on the bin that causes it to accelerate. To keep the attention on the fluid power aspects, we shall assume that drag on the bin is viscous. Sliding (i.e. Coulomb) friction can be handled quite readily in numerical analysis, but it is more difficult to handle analytically. Later we shall discuss how Coulomb friction can be included in the analysis. Applying Newton's Second Law to the soil bin yields:

$$p_L D_m \left(\frac{2N_{grbx}}{d_{cap}} \right) - c_{bin}\dot{x} = m_{bin}\ddot{x} \tag{10.13}$$

Write this in the form of two first order differential equations:

$$\frac{d\dot{x}}{dt} = \frac{1}{m_{bin}} \left(p_L D_m \left(\frac{2N_{grbx}}{d_{cap}} \right) - c_{bin}\dot{x} \right) \tag{10.14}$$

and:

$$\frac{dx}{dt} = \dot{x} \tag{10.15}$$

Thus there are three first order differential equations that can be solved numerically (Equation 10.12, 10.14, and 10.15). The input function will be $D_p(t)$ and this is assumed to be directly proportional to the voltage generated by the electronic control module.

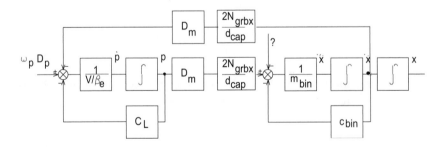

Figure 10.4: Block diagram version of the governing equations for the soil bin example.

Using block diagrams to make the relationship between a physical system and the differential equations more comprehensible was discussed in Chapter 5. The three differential equations for this system Equations 10.12, 10.14 and 10.15 have been converted into a block diagram in Figure 10.4. The first summer block represents the conservation of mass for fluid flow into the pump. The ideal pump flow less the flow into the motor and leakage is the flow into (or out of) the compliant volume represented by an equivalent bulk modulus. The integrator block converts the rate of change of pressure dp/dt into pressure and this pressure passes through the multiplier blocks D_m and $(2N_{grbx}/d_{cap})$ and exits as a force acting on the bin. The next summer block accounts for the forces on the bin. For simplification, the soil engaging tool force was neglected and Coulomb friction was replaced by viscous friction.

The quantity with the question mark entering the summer block from above could be replaced by a soil engaging tool force and Coulomb friction in the form $F_{coul}(\dot{x}/|\dot{x}|)$. The ratio of velocity and absolute value of velocity serves to switch the direction of the Coulomb force according to the direction of motion. If Coulomb force were the only motion generated resisting force, then the feedback from the velocity through the c_{bin} multiplier block would be removed. There are two integrator blocks shown in the right-hand portion of the diagram. This provision is simply to allow calculation of both velocity and displacement.

10.4.1   Example: The Soil Bin Drive

The major function of simulation is to allow the designer to examine changes in operating conditions without needing to build expensive prototypes at an early stage. Suppose the user of the soil bin had approached a designer with the need to achieve a maximum velocity of 4.5 m/s over 10 m of a 30 m track. The user indicates that the bin loaded with soil will have a mass of 2500 kg. The user would like the bin velocity during engagement of a tool not to vary by more than ±2.5%. The user indicates that there is some equipment remaining from a prior project and this should be used if possible.

After an inspection, the designer decides that coupling the variable displacement pump to the motor with flexible hoses would result in the least expensive project. After some initial examination of the existing equipment, the designer decides that the hoses would be 1.2 m long by 38 mm in diameter. The motor is a standard four pole induction motor that would have a synchronous speed of 1800 rpm at 60 Hz. Preliminary static calculations suggest that the maximum pressure will be about 7 MPa. Inspection of the manufacturer's literature indicates that the motor and pump volumetric efficiencies will be about 95% at that speed and pressure. This information allows an estimate of the system leakage coefficient to be made. Reference to Chapter 8 of this text will show that volumetric efficiency is a function of speed, so the fixed value of the leakage may not be entirely accurate.

The initial characteristics for the problem are presented in Table 10.1. A program written in Visual Basic for Applications® program within Excel®. This program solves the three first order differential equations numerically. The results for the values presented in Table 10.1 are shown in Figures 10.5

Table 10.1: Hydrostatic transmission example, initial characteristics

Characteristic	Size	Units
Pump speed, ω_p	188	rad/s
Maximum pump displacement, D_p	10.2E−06	m³/rad
Overall leakage coefficient, C_L	28.1E−12	m³/Pa · s
Oil volume (high pressure side), V	1.61E−03	m³
Equivalent bulk modulus, β_e	0.057E+09	Pa
Gearbox ratio, N_{grbx}	12.5:1	
Capstan diameter, d_{cap}	0.6	m
Motor displacement, D_m	10.2E−06	m³/rad
Bin mass, m_{bin}	2500	kg
Bin viscous drag coefficient, c_{bin}	300	N · s/m
Bin maximum velocity $\dot{x}_{max}$	4.51	m/s

to 10.7. It is obvious that the velocity varies too much during the *constant velocity* segment to be acceptable. This was really to be expected because the coupling of the pump and motor using flexible hoses introduces two problems. First, the hose itself is quite compliant and second, the volume of oil contained in it is relatively large.

Table 10.2 shows some recommended changes. The pump and motor are close coupled using steel pipes 0.2 m long. The bulk modulus of a typical hydraulic oil is about 1800 MPa. This value varies with temperature and the degree of air entrainment (see Chapter 2). In Table 10.2, the effective bulk modulus is set at 1610 MPa. This corresponds to 5% air at atmospheric pressure, rigid container (e.g. piping) volumes, and an operating pressure of 7 MPa. The results of the simulation using the values in Table 10.2 are shown in Figures 10.8 to 10.10. The variation in the constant velocity section is much less and would probably satisfy the user.

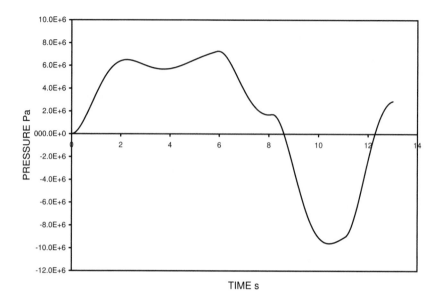

Figure 10.5: Simulated soil bin drive using flexible hoses, pressure vs. time.

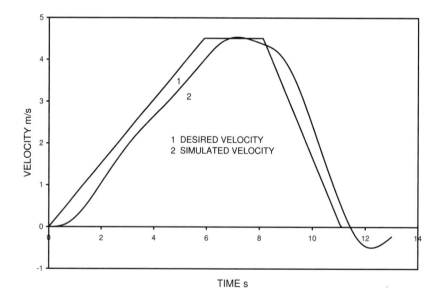

Figure 10.6: Simulated soil bin drive using flexible hoses, velocity vs. time.

Table 10.2: Hydrostatic transmission example, improved characteristics

Characteristic	Size	Units
Oil volume (high pressure side), V	0.476E−03	m^3
Equivalent bulk modulus, β_e	1.61E+09	Pa

There is little else that the designer can do to improve this design. The maximum pressure is only about 7 MPa, so the pump and motor could be reduced in size because pumps are commonly available that operate up to 20 MPa. Unfortunately, this would probably increase leakage and might not benefit the design. Examining this would be a useful student exercise. It was shown in Chapter 8 that volumetric efficiency improves with pump speed. Unfortunately induction motors have a maximum speed of 3600 rpm when operated on a 60 Hz supply. Repeating the simulation with a system based on equipment driven at 3600 rpm nominal speed would also be a useful exercise.

As a last comment, this system does not incorporate any feedback between the output velocity and the pump displacement. In this context such feedback would probably not be cost effective. In fact the velocity achieved by the bin is only 4.3 m/s. This discrepancy in velocity could easily be adjusted by altering the period of acceleration slightly. The system simulation was based on times for an ideal system with no leakage.

10.4.2 Final Comments on the Soil Bin Example

The analysis shown has made many simplifying assumptions. As indicated earlier, it is usually desirable to simplify as much as possible initially and then add complexity as the problem demands. Altering viscous drag to Coulomb friction would be a sensible modification in this design. The goal of the design was to provide a soil bin that could be used for tillage research, yet no tool force was incorporated in the design. The designer should introduce this tool force for a period after the bin reaches constant velocity and should observe how much the velocity drops as the tool engages. In a different context, it might be necessary to include the dynamics of the swash plate control in the pump. Likewise in a system where the load was much less massive, it would be necessary to include the moment of inertia of the motor as well as the load.

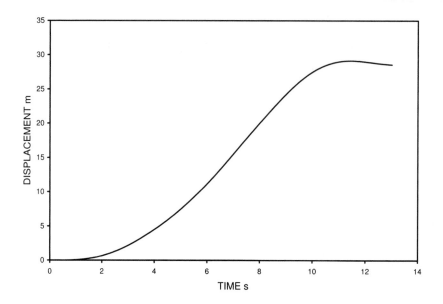

Figure 10.7: Simulated soil bin drive using flexible hoses, position vs. time.

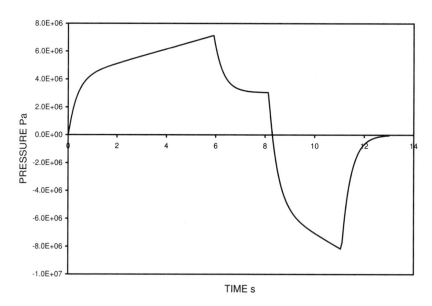

Figure 10.8: Simulated soil bin drive using rigid pipe, pressure vs. time.

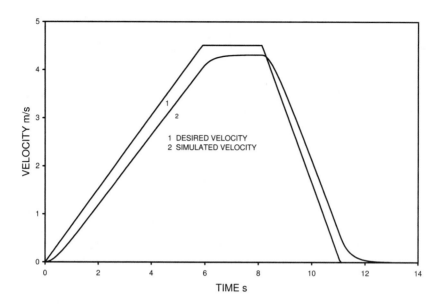

Figure 10.9: Simulated soil bin drive using rigid pipe, velocity vs. time.

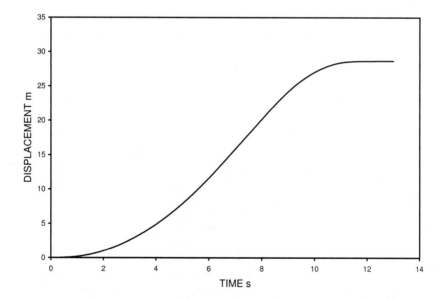

Figure 10.10: Simulated soil bin drive using rigid pipe, position vs. time.

We also indicated that the backlash in the gearbox would be ignored. In a system like the soil bin, only the forward motion dynamics are of concern, so gearbox backlash would have little or no effect. In a system that moves forward and backwards, e.g. a positioning system, the gearbox backlash effect would have to be incorporated. Another feature that might be important would be elasticity in the drive components between the motor and the bin. Simulation is always a compromise between accuracy of prediction and cost.

There is one feature of the design that should be noted by the reader. A brief discussion of the relation between torque and speed of an induction motor was given. The analysis then proceeded as if the prime mover speed was invariant and that the prime mover could both produce and absorb torque. This may not be far from true for an induction motor, but may be far from true for an internal combustion engine. If the prime mover is to provide system braking, the speed vs. torque characteristics for such operation must be determined, otherwise a potentially damaging overshoot could occur.

PROBLEMS

10.1 The two rear drive sprockets on a crawler tractor are powered with variable displacement hydraulic motors through a reduction gear set. The motors are driven by a single hydraulic pump with equal flow to each motor.

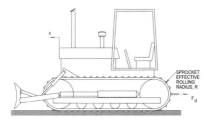

Determine the highest and lowest sprocket speeds, n_{sl} and n_{sh}, that will occur as the motor displacement advances from D_{ml} to D_{mh}. Determine the tractor drawbar pull, F_d, and the tractor power, P, that will be produced with the values for motor displacement D_{ml} and D_{mh}, and the other values given in the table.

Characteristics of a hydrostatic transmission for a crawler tractor

Characteristic	Size	Units
Pump displacement, D_p	38	mL/rev
Pump speed, n_p	1250	rpm
Pump volumetric efficiency, η_{vp}	96	%
Outlet pressure, p_s	23	MPa
Motor displacement, low, D_{ml}	230	mL/rev
Motor displacement, high, D_{mh}	550	mL/rev
Motor volumetric efficiency, η_{vm}	96	%
Motor mechanical efficiency, η_{mm}	97	%
Motor discharge pressure, p_r	230	kPa
Gear ratio, $N = n_m/n_s$	4.3:1	
Sprocket effective rolling radius, r	377	mm

10.2 The two rear drive sprockets on a crawler tractor are powered with variable displacement hydraulic motors through a reduction gear set. The motors are driven by a single hydraulic pump with equal flow to each motor.

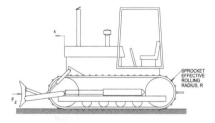

Determine the highest and lowest sprocket speeds, n_{sl} and n_{sh}, that will occur as the motor displacement advances from D_{ml} to D_{mh}. Determine the tractor dozer blade force, F_z, and the tractor power, P, that will be produced with the values for motor displacement D_{ml} and D_{mh}, and the other values given in the table.

Characteristics of a hydrostatic transmission for a crawler tractor

Characteristic	Size	Units
Pump displacement, D_p	40	mL/rev
Pump speed, n_p	1200	rpm
Pump volumetric efficiency, η_{vp}	97	%
Outlet pressure, p_s	23.5	MPa
Motor displacement low, D_{ml}	220	mL/rev
Motor displacement high, D_{mh}	570	mL/rev
Motor volumetric efficiency, η_{vm}	97	%
Motor mechanical efficiency, η_{mm}	96	%
Motor discharge pressure, p_r	240	kPa
Gear ratio, $N = n_m/n_s$	4.5:1	
Sprocket effective rolling radius, r	375	mm

10.3 The drive sprockets on a crawler tractor are powered with hydraulic motors through a reduction gear set.

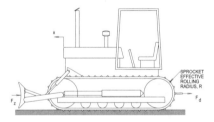

Determine the required motor displacement, D_m, to produce the given dozer force, F_z, and drawbar force, F_d. Determine the required motor flow, Q, to produce the given tractor speed, $\dot{x}$.

Characteristics of a hydrostatic transmission for a crawler tractor

Characteristic	Size	Units
Motor overall efficiency, η_{om}	93	%
Motor mechanical efficiency, η_{mm}	95	%
Motor inlet pressure, p_1	28.5	MPa
Motor return pressure, p_2	300	kPa
Gear ratio, $N = n_m/n_s$	4.1:1	
Sprocket effective rolling radius, r	373	mm
Drawbar force, F_d	15	kN
Dozer force, F_z	17	kN
Tractor speed, $\dot{x}$	5.1	km/h

REFERENCES

1. Lambeck, R. P., 1983, *Hydraulic Pumps and Motors: Selection and Application for Hydraulic Power Control Systems*, Marcel Dekker, Inc., New York, NY.

2. Steenhoek, L., Smith, R. J., Akers, A., and Chen, J., 1993, "Simulation and Validation of a Mathematical Model of a Hydrostatic Transmission", *New Achievements in Fluid Power Engineering ('93 ICFP)*, pp. 396-403.

3. Chapman, S. J., 1998, *Electrical Machinery fundamentals*, WCB/McGraw- Hill, Boston, MA.

11

PRESSURE REGULATING VALVE

11.1 PURPOSE OF VALVE

Power at any point in a hydraulic system can be determined by multiplying the fluid flow, Q, by the pressure drop, Δp, across a section of the machine. Flow is produced by an appropriate hydraulic pump. Pressure is the result of restriction in the system, caused by fluid viscosity, system geometry, and power output. Because the hydraulic pumps that are used in fluid power systems are of the positive displacement type (Chapter 8), pressure will develop up to a preset regulated value. This desired pressure value is controlled by an appropriately designed pressure regulating valve. Figure 11.1 shows a typical hydraulic power system with the pump and valve shown as items 1 and 2 respectively.

All hydraulic power systems must have at least one valve of this type. It is the function of this valve to establish the maximum pressure that will develop in the system. The pressure developed in a system will at all times be large enough to overcome circuit resistance and provide desired output power. Without a pressure regulating valve, system pressure could rise until failure of machine parts occurs.

Since the pressure regulating valve has great influence on the operation of the system, an understanding of the valve's function is important. Knowing how the valve operates provides much insight into the operation of an entire hydraulic power machine. The configuration of a typical pressure regulating valve is shown in Figure 11.2. This type of valve is easily designed and manufactured. Fortunately, the mathematical model for this

277

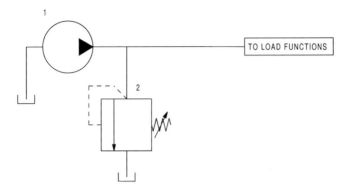

Figure 11.1: Typical fluid power system.

type of valve is reasonably simple to write and apply. Every system has basic work functions to perform. The valve model may be used to provide the desired operating characteristics for the system to which it is applied. Also, the model is often useful in examining system stability, component interaction, safety, and noise characteristics

The load functions shown in Figure 11.1 can consist of fluid power components such as valves, motors, and actuators. These components may be utilized in many configurations, depending on the desired use of the machine. In some systems, load functions are arranged in parallel branches, which can all be powered by a single pump.

Working pressure in the system will at all times develop to the necessary level required to accomplish the desired output work. The function of the pressure regulating valve, located at the bottom of Figure 11.1, is to establish the maximum pressure that can be generated in the system.

11.2 OPERATION OF VALVE

Figure 11.2 shows details for the pressure regulating valve. The valve is shown in its closed position. If the pump is started, with no load functions in operation, pump flow, Q_p, will flow into valve chamber, V_1. A rapid rise will result in pressure, p_1, which will cause the spool valve to move to the right. A force balance will then result on the valve, which consists of the pressure, p_1, multiplied by the valve end area, A, on the left and the force of the spring on the right.

As the valve moves to the right, return flow, Q_v, will begin to flow

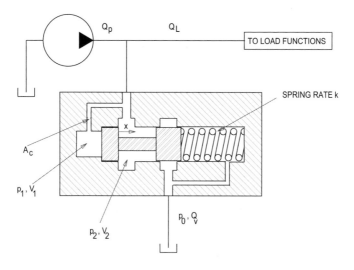

Figure 11.2: Internal details of pressure regulating valve.

to the oil reservoir through the flow area which develops. With no load functions in operation, the entire pump flow will flow to the oil reservoir. Pressure on the left end of the valve causes a force that is in balance with the force produced by the spring as it is compressed by valve motion. Proper selection of the valve diameter and the spring, allows the valve assembly to be configured to provide the desired regulated pressure, p_2. The spring can be partially compressed when installed to allow very accurate modification of the regulated system pressure.

When no load functions are in operation the valve will open to an equilibrium position and pressures, p_1 and p_2, will become equal. The entire pump flow will continue to flow back to the reservoir for this condition. When load functions are in use, load flow, Q_L, will develop. Pressure, p_2, will then be reduced to a value needed to operate the desired load functions. Pressure, p_1, will drop to the same value as pressure, p_2, and the valve will move towards the left to a partially closed position. As the use of load functions is varied the valve will continue to move to provide the desired system pressure. The pump flow will be appropriately divided between the required load flow and return flow to the reservoir. Therefore, the pressure regulating valve will at all times govern the system pressure and will provide proper flow division between the load functions and the reservoir.

The load functions for a system may consist of a single working branch or several branches. At any given time, single or multiple functions may be in use. Each function is put into operation by its own control valve, therefore, system operating pressure, p_2, may be required to change very rapidly. The valve spool will then respond as needed to meet the changing load requirements. Operation of the valve spool is very dynamic. Proper design of the valve must include the consideration of these dynamic conditions. The example which follows illustrates the general method used to provide good design.

The parameters listed in Table 11.1 are the variables and constants needed to write a basic mathematical model for the pressure regulating valve shown in Figure 11.2. The numerical values given are generally accepted average values. Many variations may be used for the numerical data given. The mathematical model which follows provides the basic equations needed to study the operation of the valve. Study of the valve allows the specific design to be customized for its intended application. The basic model does not include all of the physical effects that may apply to this type of valve. A more complete model of the valve may include additional physical effects. These effects will be introduced and discussed in Chapter 12. Experience is required to determine which physical effects are most important for modeling a particular valve configuration. Mathematical models, however, are a useful tool in determining the relative significance of applicable effects.

11.3　MATHEMATICAL MODEL OF VALVE

A mathematical model for the pressure regulating model can be established with use of the general equations listed in Chapter 4. A solution of this set of equations will then provide a preliminary description of the valve's operating characteristics.

Figure 11.3 presents a free body diagram of the valve spool with the appropriate forces noted. A summation of the forces on the valve will provide a convenient equation of motion for the valve spool. Newton's Second Law of motion as applied to the valve gives:

$$\Sigma F_x = m\ddot{x}$$

$$p_1 A - F_x - k_x x - c_x \dot{x} = m\ddot{x}$$

For the static condition, the equation may be arranged as follows:

$$p_1 = \frac{F_x + k_x x}{A}$$

Table 11.1: Characteristics for pressure regulating valve analysis

Characteristic	Size	Units
Valve diameter, d_v	0.018	m
Orifice diameter, d_o	0.001	m
Valve flow area, A_v		m^2
Oil bulk modulus, β_e	1.0E+9	Pa
Orifice discharge coefficient, C_d	0.6	
Spring preload, F	5000	N
Spring rate, k_x	210000	N/m
Valve viscous damping, c_x		$N \cdot s/m$
Valve mass, m	0.05	kg
Return pressure, p_0	0	Pa
Control pressure, p_1		Pa
Load pressure, p_2		Pa
Control flow, Q_c		m^3/s
Load flow, Q_L		m^3/s
Pump flow, Q_p		m^3/s
Valve flow, Q_v		m^3/s
Volume, V_1	5.0E−6	m^3
Volume, V_2	100E−6	m^3
Valve motion, x		m
Valve velocity, $\frac{dx}{dt}$ or $\dot{x}$		m/s
Valve acceleration, $\frac{d^2x}{dt^2}$ or $\ddot{x}$		m/s^2
Pressure rise rate, $\frac{d^2p_1}{dt^2}$ or $\dot{p}_1$		Pa/s
Pressure rise rate, $\frac{d^2p_2}{dt^2}$ or $\dot{p}_2$		Pa/s

This equation provides a convenient method for establishing the desired pressure, p_1, as a function of the spring rate and spring preload. The pressure determined with this equation will be very close to the maximum

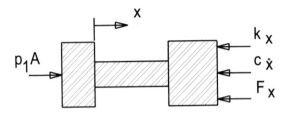

Figure 11.3: Valve spool with applied forces.

system pressure. Spool valves, of the type shown in Figure 11.2, are generally overlapped for a short length. The distance, x, that applies to the use of the equation, for the static condition, is the amount of overlap for the valve spool.

Two additional equations for the valve model may then be established with the use of the conservation of volume (strictly mass) discussed in Chapter 4, that is Equation 4.2. The equations are applied to volumes, V_1 and V_2, shown on Figure 11.2. First for volume, V_1:

$$Q_c = A\dot{x} + \frac{V_1}{\beta_e}\dot{p}_1$$

This type of expression is also referred to as a continuity equation, because it accounts for all of the fluid that is involved in a specific portion of a system. The equation accounts for the flow required to displace the spool valve motion and the flow stored due to compressibility. The final term in the equation, which involves the volume, pressure, and the pressure derivative, expresses the compressibility flow. The magnitude of this term is small, however, it is necessary because hydraulic oil is slightly compressible under pressure. The equation may be solved for the pressure derivative:

$$\dot{p}_1 = \frac{\beta_e}{V_1}(Q_c - A\dot{x})$$

Integration of this type of equation gives a very realistic measure of the pressure in a given volume [1]. The actual value for the bulk modulus term, β_e, depends on several factors some of which were discussed in Chapter 2.

Flow through the control orifice, shown in Figure 11.2 may be expressed with the use of Equation 3.19:

$$Q_c = C_d A_c \sqrt{\frac{2}{\rho}} \sqrt{p_2 - p_1}$$

The parameter, A_c, is the flow area through the valve control orifice. Its area is:

$$A_c = d_o^2 \frac{\pi}{4}$$

The size of the control orifice has considerable influence on the operation and stability of the valve. Flow, Q_c, through the orifice may be in either direction depending on the magnitudes of the pressures, p_1 and p_2. Flow into volume, V_2, may be expressed as:

$$Q_p = Q_c + Q_l + Q_v + \frac{V_2}{\beta_e} \dot{p}_2$$

Valve flow, Q_v, also may be expressed with the use of Equation 3.19 as:

$$Q_v = C_d A_v \sqrt{\frac{2}{\rho}} \sqrt{p_2 - p_0}$$

The flow area, A_v, for the type of valve shown in Figure 11.2 is dependent on the valve spool motion, x. Because the spool valve is overlapped, the flow area is zero until the valve has moved the full overlap distance. As the valve continues to move flow area, A_v, develops as a function of the valve geometry. Flow area, as a function of geometry, may be described with the valve area gradient, as expressed in Equation 3.20. The gradient, which is generally described with the parameter, w, has units of in.^{2}/in. or m^2/m. The gradient may be linear or nonlinear depending on valve geometry. Therefore:

$$A_v = wx$$

11.4   EFFECT OF DAMPING

Damping is present in a fluid power system, such as that described, because of several physical effects. These effects are difficult to fully define. General descriptions, however, can be found in existing literature [1-3].

The major contributors to damping in a hydraulic system are Coulomb friction, viscous friction, fluid flow forces, and hysteresis in parts that deform under loads. Damping exists in a system for each component as well as for the overall system. Because of the variety of phenomena that may cause damping, damping is difficult to express analytically in a mathematical model for a fluid power system. Incorporating damping, however, is most important for describing the realistic motion of a system. A concept derived in classical mechanics can be incorporated in the model. Typically, fluid power systems contain one or more spring-mass-damper systems. Although a simplification, components such as valve spools can often be modeled quite adequately by postulating that only viscous damping is present. An analysis of a spring-mass-damper system with viscous damping was undertaken in Chapter 5 where it was shown that the dimensionless term, the damping coefficient ζ, had a value of 1 on the boundary between an underdamped (i.e., oscillatory) system and an overdamped system. A value of $\zeta = 1$ is defined as critical damping.

Actual spring-mass systems will typically have a fraction of the critical damping value. An estimation of the percentage to apply has been gained from actual laboratory test results. Measured variables such as pressure and mass motion may be displayed vs. the independent variable, time. When the operation of the system is stopped, the variables will display a decay curve such as that described in classical vibrations texts. The shape of these curves indicates that a mass that is free to move, such as a valve spool in a housing bore, will typically show a percentage of damping that will be approximately 50% of critical. For masses that most move against much friction, such as that provided by oil seals, the damping value will approach 100% of critical.

For a spring-mass-damper combination, damping can be expressed as [4]:

$$c_x = \sqrt{k_x m} \qquad (11.1)$$

The justification for this specific expression may be seen by examining Equation 5.24 when $\zeta = 0.5$.

The damping coefficient, c_x, derived by this method is multiplied by the velocity of the mass to which it applies. The resulting force opposes motion of the mass as described in the equation for valve motion.

Use of Equation 11.1 then allows the percentage of critical damping to be easily varied to study its influence on the system. Some systems require the addition of damping to provide desired operational stability. Damping can be added through a variety of methods. The most common methods employ the use of orifices, dashpots, and friction. Various arrangements will allow the addition of damping which can be made proportional to the velocity of a moving mass. Investigation on the influence of the added damping can then be accomplished with the equation of motion given above. Use of a sharp edged orifice in a dashpot arrangement will provide velocity squared damping. The velocity term in the equation of motion would then be squared to provide a very aggressive type of damping. The use of sliding (e.g., Coulomb) friction provides the easiest means of adding damping to a system. This type of damping is difficult to evaluate mathematically, however, it can be tailored to a specific application through trial and error means.

11.4.1   Example: Solution of Model

A solution for the general mathematical model developed in Section 11.3 follows with use of information listed in Table 11.1. Other specific information, needed for the solution, is noted in the example.

Flow from the pump, Q_p, which serves as the input to the system, was increased linearly from 0 L/s at time zero to 1.0 L/s at 0.05 s. Flow remained at 1.0 L/s after time 0.05 s. Flow area, A_v, for the spool valve is 0 mm² for the valve overlap distance of 0.635 mm. Area increased linearly from 0 mm² at valve opening distance of 0.635 mm to 121.32 mm² at the maximum spool position. Mathematically, the valve flow area may be described in m² with the expression:

$$A_v = 0.06367x - 0.0000404 \qquad \text{where} \qquad 0.000635 < x < 0.00254$$

Values for valve spool position must be in units of meters for this expression. Valve spool motion must be constrained between the values of 0 and 0.00254 m, when the equations for the pressure regulating valve are executed.

The mathematical model, for the pressure regulating valve, may be solved with any simulation program capable of integrating the differential equations. For a correct solution the motion, x, of the valve spool must be limited between zero and 2.54 mm. The direction of flow through the control orifice, Q_c, depends on the magnitude of the pressures, p_1 and p_2. Therefore, this condition must also be accounted for in the solution method.

Some oscillations are present in the variables shown in Figures 11.4 to 11.6. These oscillations are predominantly due to the influence of the oil compliance, which results from the oil compressibility factor (also called bulk modulus in Chapter 2), β_e [1, 2]. The spool valve and its control spring function as a vibrating spring mass system, which also contributes to the oscillations.

All valves of this type have a variety of conditions present which contribute damping [2]. Damping, which will be revisited in Chapter 12, has considerable influence on the operation of the valve and the behavior of variables.

Pressure, p_1, develops very rapidly because the control volume, V_1, is small. This exact volume size is somewhat speculative, however, it appears reasonable to end the volume at the control orifice. Valve entry volumes such as volume, V_2, generally include some of the flow line volume.

With no load flow, Q_L, the valve will open far enough to return all of the pump flow, Q_p, to the system reservoir. The model is useful in determining the basic design of the valve. The model may then be used

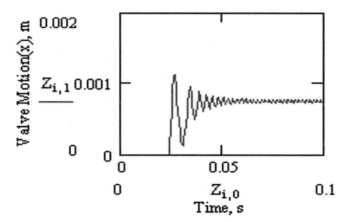

Figure 11.4: Valve spool motion vs. time.

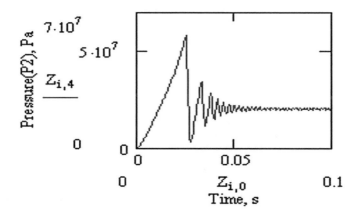

Figure 11.5: Regulated pressure, p_2, vs. time.

to explore other system conditions such as variations in load flow, pump flow, damping factors, valve spool flow forces, and oil compressibility. The influence of some of these effects are examined elsewhere in the text.

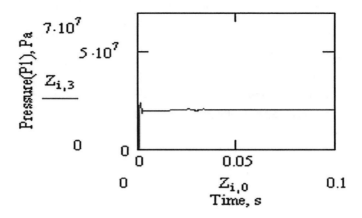

Figure 11.6: Controlled pressure, p_1, vs. time.

PROBLEMS

11.1 A pressure regulating valve is used to set a system pressure, p, to control the fluid power motor as shown in the figure.

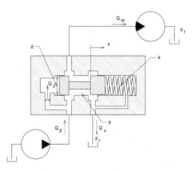

Determine the valve spool position, x, that is required to produce the given conditions. Oil flow, Q_v flows to the tank around the complete circumference of the spool valve. Determine the initial deflection, δ, that is required for the spring to produce a spring preload that will be consistent with the given system operating conditions.

Pressure regulating valve in steady state

Characteristic	Size	Units
Motor speed, n	1000	rpm
Motor torque, T_m	125	N · m
Motor displacement, D_m	50.0	mL/rev
Motor mechanical efficiency, η_{mm}	94.0	%
Motor volumetric efficiency, η_{vm}	95.0	%
Pump flow, Q_p	0.94	L/s
Orifice flow, Q_c	0.0	L/s
System return pressure, p_r	250	kPa
Valve overlap, U	0.30	mm
Valve flow coefficient, C_d	0.62	
Valve diameter, d	12.5	mm
Spring rate, k	167	kN/m
Oil density, ρ	850	N · s^2/m^4

11.2 A hydraulic lift is used to raise a 4200 lb$_m$ mass vehicle as shown in the figure.

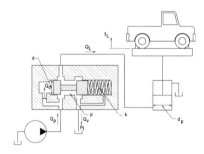

Determine the valve spool position, x, that is required to produce the given conditions. Oil flow, Q_v flows to the tank around the complete circumference of the spool valve. Determine the initial deflection, δ, that is required for the spring to produce a spring preload that will be consistent with the pressure required to lift the vehicle.

Pressure regulating valve used to control a vehicle hoist

Characteristic	Size	Units
Lift distance, ℓ_L	51.0	in.
Lift time, Δt	5.0	s
Piston diameter, d_p	3.55	in.
Pump flow, Q_{pi}	27.0	gpm
Orifice flow, Q_c	0.0	gpm
Tank pressure, p_t	0.0	$lb_f/in.^2$
Valve overlap, U	0.01	in.
Valve flow coefficient, C_d	0.6	
Valve diameter, d	0.4	in.
Spring rate, k	350	$lb_f/in.$
Oil density, ρ	0.078	$lb_f \cdot s^2/in.^4$

11.3 A pressure regulating valve is used to set a system pressure, p, to control the fluid power motor as shown in the figure.

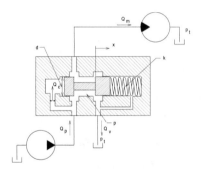

Determine the valve spool position, x, that is required to produce the given conditions. Oil flow, Q_v flows to the tank around the complete circumference of the spool valve. Determine the initial preload, F, that is required for the spring to produce the given system operating conditions.

Pressure regulating valve in steady state

Characteristic	Size	Units
Motor speed, n_m	857	rpm
Motor torque, T_m	215	N · m
Motor displacement, D_m	82.0	mL/rev
Motor mechanical efficiency, η_{mm}	97.0	%
Motor volumetric efficiency, η_{vm}	95.0	%
Pump speed, n_p	1800	L/s
Pump displacement, D_p	45.0	mL/rev
Pump volumetric efficiency, η_{pv}	96.0	%
Orifice flow, Q_c	0.0	L/s
System return pressure, p_t	300	kPa
Valve overlap, U	0.32	mm
Valve flow coefficient, C_d	0.6	
Valve diameter, d	12.7	mm
Spring rate, k	170	kN/m
Oil density, ρ	850	N · s^2/m^4

REFERENCES

1. Blackburn, J. F., Reethof, G., and Shearer, J. L., 1960, *Fluid Power Control*, The M.I.T. Press, Cambridge, MA.

2. Burton, R., 1958. *Vibration and Impact*, Addison-Wesley, Reading, MA.

3. Merritt, H. E., 1967, *Hydraulic Control Systems*, John Wiley & Sons, New York, NY.

4. Gassman, M. P., 1997, "Mathematical Analysis of a Fluid Flow Control Valve", SAE Paper 971579, Peoria, IL.

12

VALVE MODEL EXPANSION

12.1 BASIC VALVE MODEL

A mathematical model for a typical pressure-regulating valve was developed in Chapter 11, Section 11.3. This model demonstrates the general operation of the valve and can be used to establish the desired magnitude of basic operating parameters. Generally the basic model is used to establish reasonable values for the geometry of the valve parts. Values for these parameters are listed in Table 11.1. Knowledge of desired valve performance serves as a guide in establishing parameter values. Many variations are possible, in the listed parameters, and in establishing desired valve configuration. Results of the solution of the model equations, as displayed in Figures 11.4 to 11.6, indicate that the model provides a reasonable simulation of the valve.

The model includes only the basic parameters needed to describe valve operation. Several other parameters may be added to the model to provide for an investigation of their influence on the operation of the valve. The additional parameters may be evaluated individually or in any desired combination. If the inclusion of a particular parameter appears to have significant influence on the performance of the model, investigation of the parameter over its expected range of values is usually desirable.

Many fluid power systems have a dominant characteristic. Typically, the effect of a mass, a spring, or the oil viscosity may exhibit a dominant role in the machine's operation [1]. Repeated solution of the mathematical model with variation in applicable parameters will reveal which characteristics are of greatest importance.

In addition to parameters listed in Tables 11.1 and 12.1, other major physical conditions that may have an effect on the model's performance are damping, flow forces, and temperature. This chapter will examine the influence of several important parameters. The varieties of conditions that can be imposed on a model are unlimited. Even preliminary results of a model's solution, however, will quickly reveal important operational characteristics for a system. Before proceeding with the addition of new terms in the modeling equations, it is important to note what has already been learned from the results displayed in Chapter 11. The main purpose of mathematical modeling is to gain knowledge regarding trends in the operation of a system. Not all numerical values will agree exactly with results that may be obtained in laboratory operation of the system under study. Simulated values, however, will usually be accurate enough to provide much insight into the operation of a system.

Fluid power systems are said to be in steady state operation when the variables cease to vary with time. Fluid systems that are used to transfer power, however, contain such elements as springs and masses. The physical operation of these parts prevents the system from reaching steady state conditions. This is particularly true for masses that are in contact with a spring. Hydraulic oil is slightly compressible; therefore, the entire system tends to behave as a large spring. Hydraulic systems, therefore, have an overall stiffness or compliance [2].

Fluid power systems are also subject to a variety of disturbances. Disturbances can include opening and closing of control valves, operation of accumulators, and changes in external loads. The fluid pumps used to power systems contain moving elements such as pistons, vanes, or gear teeth. These moving elements impose intermittent disturbing forcing functions on the fluid stream.

Peak values of variables, such as those plotted in Chapter 11, will generally vary several percent from measured values obtained in laboratory tests. These peaks determined with mathematical simulation are generally accurate enough for most engineering purposes. Plotted variables at the extreme right of the plots will agree very closely with laboratory values, because enough time has passed to correct for the effect of start-up assumptions.

Generally fluid pressure is the most significant parameter in a system. The entire machine responds to the value of this parameter. Knowledge regarding expected pressure values in a fluid power system is also necessary to provide for proper design of parts for safe stress levels and safety.

The valve spool mass resting on the coil spring essentially behaves as a classical spring-mass system. Free oscillation of the valve spool mass on the spring will occur at the natural frequency of the spring-mass system. Results developed from the study of classical vibrations gives this result as:

$$f_n = \frac{1}{2\pi}\sqrt{\frac{k}{m}} \qquad (12.1)$$

Substituting the parameters for the example in Chapter 11 yields:

$$f_n = \frac{1}{2\pi} \sqrt{\frac{210000}{0.05}} = 326.2 \text{ Hz}$$

In addition to the natural frequency of elements, such as the spring-mass combination, a system will have an overall natural frequency [1, 2]. Therefore, oscillation frequencies, such as those displayed in the plots displayed in Section 11.3, will include the overall effect of all the parameters included in the system model.

It would be somewhat impractical to attempt to include all possible physical effects in a system model. The sections that follow will examine some of the most important effects that can be expected in a fluid power model.

12.2 MODEL EXPANSION

Consider a complete circuit shown in Figure 12.1. The analysis for the pump (component 1) and pressure regulating valve (component 2) operating with a mathematically defined pump flow variation and no load flow was presented in Chapter 11. The figure shows the addition of a fluid power cylinder and an appropriate control valve. The cylinder is used to raise the load mass m_L. The parameters needed to expand the model are shown in Table 12.1 Flow area A_v through the control valve, shown as component number 3, was input as a time dependent function. This area was held at 0 from time 0 to 0.035 s. The area was then increased linearly from 0 to the maximum value shown in Table 12.1 in the time interval from 0.035 to 0.085 s. After time 0.085 s, the maximum value was maintained.

Oil flow Q_p from the pump was established as a time dependent function in Section 11.3. This same function was applied to the model established for Figure 12.1. In the analysis considered in Section 11.3, all of the pump oil flow passed through the regulating valve and returned to the oil tank. When a load is added to a pressure-regulating valve, all of the pump oil flow will flow to the work circuit. Part of the flow will begin to pass through the regulating valve and back to the oil tank when the system load is large enough to develop the regulated pressure. The value used for the load mass m_L, as shown in Table 11.1, was selected to develop a static pressure of about 70% of the regulated pressure. Therefore, all of the pump flow was directed to the cylinder.

The equations that follow are then necessary to complete the model for the entire system. The equation of motion for the cylinder piston and load can be established from the free body diagram shown in Figure 12.2 with

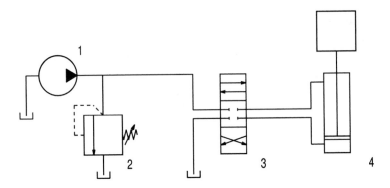

Figure 12.1: Pressure-regulating valve with control valve and cylinder load.

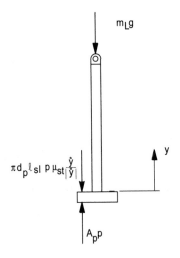

Figure 12.2: Forces on cylinder piston.

Table 12.1: Characteristics for control valve and hydraulic cylinder

Characteristic	Size	Units
Load mass, m_L	2500	kg
Acceleration of gravity, g	9.81	m/s^2
Piston diameter, d_p	0.048	m
Coefficient of static friction, μ_{st}	0.05	
Seal width, ℓ_{sl}	0.0025	m
Valve flow area, A_v	3.0E$-$6	m^2
Initial volume, V_o	5.0E$-$5	m^3
Flow coefficient, C_d	0.6	
Fluid density, ρ	832	kg/m^3
Cylinder pressure, p		N/m^2
Load flow, Q_L		m^3/s
Pump flow, Q_p		m^3/s
Piston motion, y		m
Piston velocity, $\frac{dy}{dt}$ or $\dot{y}$		m/s
Piston acceleration, $\frac{d^2y}{dt^2}$ or $\ddot{y}$		m/s^2
Pressure rise rate, $\frac{dp}{dt}$ or $\dot{p}$		Pa/s

use of Newton's Second Law:

$$A_p p - m_L g - \pi d_p \ell_{sl} p \mu_{st} \frac{\dot{y}}{|\dot{y}|} = m_L \ddot{y}$$

To determine load piston velocity $\dot{y}$ and position y the equation may be arranged as:

$$\ddot{y} = \left(A_p p - m_L g - \pi d_p \ell_{sl} p \mu_{st} \frac{\dot{y}}{|\dot{y}|} \right) / m_L$$

The piston work area A_p, required in the above equations, is equal to:

$$A_p = d_p^2 \frac{\pi}{4}$$

A value for the damping coefficient is difficult to determine for a cylinder piston. It is well known from laboratory work, however, that a friction force exists at the piston and rod seals. In the above equation the friction has been modeled as a single value that incorporates piston diameter d_p, seal width ℓ_{sl}, load pressure p, and seal friction μ. The friction force always opposes motion of the piston. Therefore, the direction of the friction force must be corrected with the ratio of piston velocity over the absolute velocity $\dot{y}/|\dot{y}|$.

Values for the cylinder pressure p can be established with the use of the continuity of flow principle. Flow continuity, as it applies to the cylinder, includes flow into and out of the cylinder barrel volumes. Also, the influence of fluid compressibility and the effect of moving parts must be included. Flow continuity applied to cylinder volume below the piston may be written as:

$$Q_i = -A_p\dot{y} + \frac{V}{\beta_e}\dot{p}$$

The volume V is a variable and is expressed as:

$$V = A_p y + V_o$$

Where flow Q_i is flow through the control valve and may be expressed as:

$$Q_i = C_d A_v \sqrt{\frac{2}{\rho}} \sqrt{p - p_2}$$

The continuity equation given above may be arranged as:

$$\dot{p} = \frac{\beta_e}{V}\left(Q_i - A\dot{y}\right)$$

This equation is integrated to establish a value of the cylinder pressure p at any time during the solution. Working pressure in the system will at all times develop to the necessary level required to accomplish the desired output work. The function of the pressure-regulating valve, discussed in Section 11.3, is to establish the maximum pressure that can be generated in the system.

The model that has been defined provides the basic equations needed to study the operation of the system. Study of the system allows the specific design to be customized for its intended application.

12.2.1 Example: Solution of Model

A solution for the general mathematical model developed in Section 12.2 follows with use of information listed in Tables 11.1 and 12.1. Every hydraulic system simulation must address specific goals. Equations and parameters must then be consistent with achieving the desired goals. The

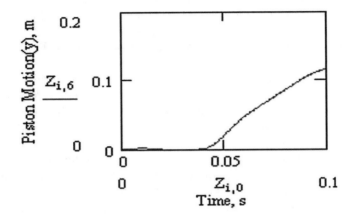

Figure 12.3: Cylinder piston motion, y, vs. time.

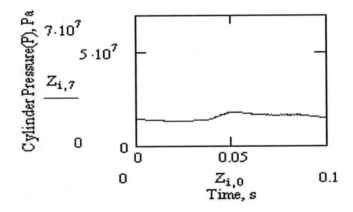

Figure 12.4: Cylinder working pressure, p, vs. time.

mathematical models that result for a specific purpose may then be altered to examine other effects as needed. Other specific information, needed for the solution, is noted in the example.

Information regarding the operation of the pressure-regulating valve is covered in Chapter 11. Figures 11.4 to 11.6 show the valve motion x and internal valve pressures p_1 and p_2. Figures 12.3 and 12.4 show the time dependent information on the cylinder piston motion y and cylinder working pressure p.

The type of mathematical model developed in this chapter can be solved with a variety of computer software programs. Solution methods may require the use of mathematical modeling equations or they may allow a graphical method. Graphical methods recognize the existence of the appropriate equations that are needed to describe the system. In general, very little variation occurs in results with the use of different programs.

As noted in Section 11.3, the control valve for the cylinder begins to open at time 0.035 s. Therefore, the cylinder piston begins to move upward at that time. The load weight is resting on the cylinder piston at time 0; therefore, the initial pressure in the cylinder is equal to the value of the load weight divided by the piston area. As the piston begins to accelerate, the cylinder pressure increases. As the cylinder piston approaches steady state motion, the pressure approaches the initial static value.

12.3   AN ASSESSMENT OF MODELING

The successful design of a machine usually requires an iterative approach. The primary purpose of the mathematical models that has been developed in Chapters 11 and 12 is the ability to easily vary the system parameter values. The machine can then be tailored to meet the need for which it is intended.

Models can easily be expanded to include addition components and circuit branches. As additional circuit branches are added, the manner in which available pump flow is divided may be of importance to the solution. Methods of determining flow division are discussed in Chapter 13. Dynamic system effects may be analyzed in the same model with flow division details.

REFERENCES

1. Merritt, H. E., 1967, *Hydraulic Control Systems*, John Wiley & Sons, New York, NY.

2. Blackburn, J. F., Reethof, G., and Shearer, J. L., 1960, *Fluid Power Control*, The M.I.T. Press, Cambridge, MA.

13

FLOW DIVISION

13.1 INTRODUCTION

The success of agricultural, manufacturing, transportation, military and construction machinery owes a great deal to the effective use of fluid power. Most fluid power systems are configured with a positive displacement fluid pump that is large enough to meet the flow requirements of many work circuits. Different work functions require a variety of fluid flow and pressure values to provide the desired operation. System branches, therefore, must include specialized flow and pressure regulating valves. The development of a mathematical model of a fluid flow control valve follows.

13.2 THE HYDRAULIC OHM METHOD

Design of a fluid power system can be improved with the use of mathematical simulation. Numerous approaches for modeling fluid power systems and components can be found in the literature. Analysis of a fluid power system may cover flow distribution, component operation, or a combination of both. Most equations useful to fluid power analysis are derived from the law of conservation of energy, the continuity principle, and Newton's Second Law. Many equations used to calculate flow in circuits involve the use of empirical expressions or laboratory derived flow coefficients. Therefore, when two or more circuits are used simultaneously, the continuity principle may not appear to be obeyed exactly, because of the use of such empirical coefficients.

In order to model a flow regulator valve, a good means must also be used to define how the available flow divides in the active circuit branches.

Usually this is done by writing a set of equations that can be solved by an iterative method to determine desired pressure and flow values. These iterative methods work well for steady state flow conditions, however, they are difficult to apply to unsteady state situations. Esposito demonstrated an excellent means of satisfying flow continuity in networks with the use of direct current (DC) electric hydraulic analogy methods [1]. In this method, fluid pressure, flow, and flow resistance are analogous to electric voltage, current, and resistance. The method uses the Lohm Law principle, also referred to as the hydraulic ohm, to describe fluid flow resistance as defined by the Lee Company [2]. The symbol R is generally assigned to flow resistance.

13.3   BRIEF REVIEW OF DC ELECTRICAL CIRCUIT ANALYSIS

Before we explain how fluid power networks may be analyzed using resistance network methods, it may be useful to review electric circuit analysis. We shall only consider resistances and motors. Incidentally, hydraulic capacitance and inductance do have a role in fluid power circuits as will be seen in Chapter 14.

Consider the circuit shown in Figure 13.1. First observe Kirchhoff's

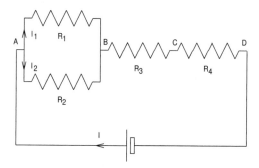

Figure 13.1: Simple electrical network with loops.

laws [3]. The algebraic sum of the currents at the junction of two or more conductors is zero. In symbolic form this is written:

$$\sum_{n=1}^{n=r} I_n = 0 \qquad (13.1)$$

The algebraic sum of potential around a closed loop is zero. That is:

$$\sum_{n=1}^{n=r} E_n = 0 \tag{13.2}$$

For electrical circuits, the relation between potential (E), current (I), and resistance (R) is:

$$E = RI \tag{13.3}$$

13.3.1 Methods of Solving DC Networks

If we apply Kirchhoff's current rule to junction A in Figure 13.1, then:

$$
\begin{aligned}
I + (-I_1) + (-I_2) &= 0 \\
I - I_1 - I_2 &= 0
\end{aligned}
\tag{13.4}
$$

There are several ways in which the circuit can be analyzed. One can apply Kirchhoff's rules and derive a set of linear simultaneous equations in the E and I variables. This might be a good procedure for an electrical circuit, but it is usually less useful for a fluid power circuit because the relation between potential (pressure) and current (flow) is generally nonlinear. Consequently, the set of equations formed is nonlinear and not easy to solve using simple techniques.

An alternative solution to the problem, which can often be applied to fluid power circuits, is to consolidate individual resistances into one equivalent resistance. This strategy does not eliminate the need to work with nonlinear equations, but the solution of the dynamics of fluid power flow problems in networks usually requires an iterative solution using a computer.

It may be found that several consolidation steps are required. For the example shown in Figure 13.1 common sense or the application of Equation 13.2 shows that the potential across resistances R_1 and R_2 must be equal:

$$E_{AB} = R_1 I_1 \quad \text{or} \quad I_1 = \frac{E_{AB}}{R_1} \tag{13.5}$$

and

$$E_{AB} = R_2 I_2 \quad \text{or} \quad I_2 = \frac{E_{AB}}{R_2} \tag{13.6}$$

If we write the equivalence of R_1 and R_2 in the parallel configuration between A and C as R_5, then:

$$E_{AB} = R_5 I \tag{13.7}$$

From Equation 13.4:

$$I = I_1 + I_2$$

So:

$$E_{AB} = R_5(I_1 + I_2)$$

Thus using Equations 13.5 and 13.6:

$$E_{AB} = R_5 \left(\frac{E_{AB}}{R_1} + \frac{E_{AB}}{R_2} \right)$$

yielding the well known result:

$$\frac{1}{R_5} = \frac{1}{R_1} + \frac{1}{R_2} \tag{13.8}$$

Although most readers will recognize that the equivalent resistance for a series of electrical resistances will simply be the sum of the set, we shall prove this formally because the strategy will be used again for a series of fluid power resistances. It will be seen shortly that a simple sum is not correct for a series of resistances in a fluid power circuit because the $p = f(R, Q)$ relation is nonlinear.

The current through the equivalent resistances, R_5, and resistances R_3 and R_4 is I. Now apply the Kirchhoff potential rule (Equation 13.2) to the loop ABCD then:

$$E_{AD} = E_{AB} + E_{BC} + E_{CD}$$

Using R_6 for the consolidated values of R_5, R_3, and R_4, the last expression can be written:

$$R_6 I = R_5 I + R_3 I + R_4 I$$

so:

$$R_6 = R_5 + R_3 + R_4 \tag{13.9}$$

The result that was expected.

We now have all the information necessary to find all potentials and currents in the circuit. We can start by using the result of Equation 13.9 to determine the current, I, provided by the voltage source:

$$I = \frac{E}{R_6} \tag{13.10}$$

Because the value R_5 is known from Equation 13.8, the value of E_{BC} can be calculated from Equation 13.7. Then the values of I_1 and I_2 can be calculated from Equations 13.5 and 13.6.

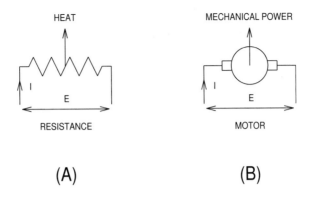

Figure 13.2: Forms of electrical power converters, the resistor and the motor.

13.3.2   Motor and Resistance Equivalence

Consider the two devices shown in Figures 13.2(A) and 13.2(B). In Figure 13.2(A), the electrical component is a resistance, so the effect of current flowing across a potential drop is to dissipate electrical energy as heat. On the other hand, in Figure 13.2(B) the electrical energy is converted into mechanical power. If an observer were to place an ammeter and a voltmeter in the circuit to measure current and potential, this observer could not tell what form of energy was being produced from the devices. As far as circuit analysis is concerned, it will be useful to treat the motor as if it were an equivalent resistance. In practice, electric motors come in several configurations and the relation between power output and current and potential input may be quite complex. For our purposes we can limit discussion to a motor with a permanent magnet field. If this motor is considered to have no friction drag or ohmic losses, then the motor performance can be described by the relations for torque:

$$T = K_1 I$$

and speed:

$$\omega = K_2 E$$

The power developed by the motor will be:

$$P = T\omega$$

(Incidentally this shows that $K_1 K_2 = 1$ because $P = T\omega = EI$.) Suppose the motor is driving an idealized winch with a constant load and no losses.

In this situation, the torque on the winch will be independent of speed, and the current taken by the motor will be constant. Thus:

$$P = I_A E \text{ where } I_A \text{ is constant}$$

For DC electrical circuits, we can write:

$$P = IE = I^2 R = \frac{E^2}{R} \tag{13.11}$$

Using Equation 13.11:

$$
\begin{aligned}
I_A^2 R &= I_A E \\
R &= \frac{E}{I_A}
\end{aligned}
\tag{13.12}
$$

Thus the equivalent resistance for the permanent magnet motor used with this specific load is not fixed, but a linear function of the applied potential. The electrical network could still be solved using the resistance consolidation method, but the equations would be nonlinear. In a simple situation like this, the equations could be still be solved explicitly.

A more general method, and one more applicable to fluid power circuits, would be to start with an initial estimate of the potential across the motor. This estimate would be used to estimate the motor equivalent resistance R. One iteration of calculation would be performed and a better estimate of the motor potential drop would be obtained. This new estimate would be used to calculate a new estimate of the equivalent motor resistance. The iterations could stop when the relative change in potential reached an acceptable value. A worked example will not be presented because this is not an electrical engineering text and we are only using the electrical analogy to lead into methods for fluid power circuit calculations.

13.4   FLUID POWER CIRCUIT BASIC RELATIONSHIPS

If we initially ignore devices that perform mechanical work, there are three classes of component that may be encountered in fluid power circuits:

PIPES   ORIFICES   FITTINGS

It should be noted that valves are considered a subclass of orifices because the flow in valves is generally treated as orifice flow (Chapter 7).

The relationship used for pipe flow is:

$$h = f\frac{\ell v^2}{2gd}$$

Using:

$$p = \rho gh \quad \text{and} \quad Q = Av$$

yields:

$$p = f\frac{\rho \ell}{2dA^2}Q^2 \tag{13.13}$$

The relationship for an orifice is:

$$Q = C_d A_o \sqrt{\frac{2p}{\rho}}$$

which can be rearranged to:

$$p = \frac{\rho}{2(C_d A_o)^2}Q^2 \tag{13.14}$$

The relationship for a fitting is:

$$h = K_h\frac{v^2}{2g}$$

or:

$$p = \frac{\rho K_h}{2A^2}Q^2 \tag{13.15}$$

It should be noted that the area term, A, should be appropriate to the component being considered. Use will follow standard fluid mechanics practice.

It will be observed that the relationships in Equations 13.13 to 13.15 are all of the form $p = f(Q^2)$. The relationship between pressure (potential), flow (current), and resistance for fluid power circuits is usually written:

$$\sqrt{\Delta p} = RQ \tag{13.16}$$

Thus we can write expressions for fluid flow components as:

Pipes

$$R = \sqrt{f\frac{\rho}{2}\frac{\ell}{d}\frac{1}{A^2}}$$

$$\tag{13.17}$$

Orifices
$$R = \sqrt{\frac{\rho}{2}} \frac{1}{C_d A}$$

(13.18)

Fittings
$$R = \sqrt{K_h \frac{\rho}{2} \frac{1}{A^2}}$$

(13.19)

If fluid flow is known to be laminar, then a relation $p \propto Q$ may be more appropriate. Fortunately turbulent and laminar flow conditions may be combined in the same analysis by using an iterative solution. This is done by making initial estimates of resistance terms in the form shown in Equation 13.16, solving the system, and then applying corrections to the R terms before performing more iterations.

Actuator and motor components also have pressure drops across them and flows through them. As with the electric motors discussed in Section 13.3.2, an equivalent resistance may be introduced to accommodate these components. First consider an actuator that has no seal friction or leakage. We can write:

$$Q = A\dot{y}$$

rearrange this as:

$$1 = \frac{Q}{A\dot{y}}$$

(13.20)

There are many options for obtaining an equivalent resistance, but the most obvious seems to be to take Equation 13.20 and multiply both sides by $\sqrt{\Delta p}$, yielding:

$$\sqrt{\Delta p} = \frac{\sqrt{\Delta p}}{A\dot{y}} Q$$

(13.21)

Thus the equivalent resistance term for an ideal actuator can be written:

$$R = \frac{\sqrt{\Delta p}}{A\dot{y}}$$

(13.22)

Whether to use $A\dot{y}$ or Q in the denominator of Equation 13.22 will depend on context. Obviously the equation is derived from terms that may vary during an iterative solution and the resistance must be updated after each iteration.

Similar reasoning for an ideal motor leads to the expression:

$$R = \frac{\sqrt{\Delta p}}{\omega D_m} \tag{13.23}$$

Real actuators and motors have torque and volumetric efficiencies that are less than unity. If more accurate estimates of flow division are required, it may be necessary to modify the actuator or motor equivalent resistance by introducing leakage and friction effects.

13.5 CONSOLIDATION OF FLUID POWER RESISTANCES

Consider the set of parallel resistances shown in Figure 13.3(A). Working

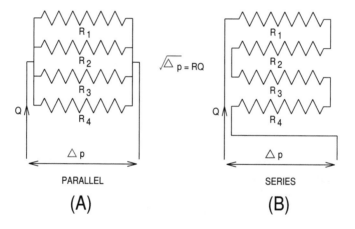

Figure 13.3: Parallel and series fluid flow resistance networks.

on the basis that the pressure across each resistance is the same and that the consolidated value of the resistances is R_{PAR} then:

$$Q = \frac{\sqrt{\Delta p}}{R_{PAR}}, \ Q_1 = \frac{\sqrt{\Delta p}}{R_1}, \ Q_2 = \frac{\sqrt{\Delta p}}{R_2}, \ Q_3 = \frac{\sqrt{\Delta p}}{R_3}, \ \text{and } Q_4 = \frac{\sqrt{\Delta p}}{R_3}$$

and that the sum of algebraic flows into a junction is zero so:

$$Q - Q_1 - Q_2 - Q_3 - Q_4 = 0$$

thus:

$$\frac{\sqrt{\Delta p}}{R_{PAR}} = \frac{\sqrt{\Delta p}}{R_1} + \frac{\sqrt{\Delta p}}{R_2} + \frac{\sqrt{\Delta p}}{R_3} + \frac{\sqrt{\Delta p}}{R_4}$$

$$\frac{1}{R_{PAR}} = \frac{1}{R_1} + \frac{1}{R_2} + \frac{1}{R_3} + \frac{1}{R_4} \qquad (13.24)$$

Now consider the set of series resistances shown in Figure 13.3(B). Working on the basis that the flow through each resistance is the same and that the consolidated value of the resistances is R_{SER} then:

$$\sqrt{\Delta p} = R_{SER}Q,$$

$$\sqrt{\Delta p_1} = R_1 Q, \ \sqrt{\Delta p_2} = R_2 Q, \ \sqrt{\Delta p_3} = R_3 Q, \text{ and } \sqrt{\Delta p_4} = R_4 Q$$

The pressure across the complete series chain is the sum of the individual pressures across each resistance. Note however, that the previous expressions must be squared to obtain these pressures so:

$$\Delta p = \Delta p_1 + \Delta p_2 + \Delta p_3 + \Delta p_4$$

$$R_{SER}^2 Q^2 = R_1^2 Q^2 + R_2^2 Q^2 + R_3^2 Q^2 + R_4^2 Q^2$$

thus:

$$R_{SER} = \sqrt{R_1^2 + R_2^2 + R_3^2 + R_4^2} \qquad (13.25)$$

Note, as indicated in Section 13.3.1, this is not the same result obtained for a series of electrical resistances in a DC circuit. It should be obvious that the procedures outlined can be performed for any number of resistances.

13.5.1 Example: Invariant Resistances

Use of the hydraulic ohm to determine flow division is illustrated in the following elementary example shown in Figure 13.4. The flow restriction in each of the two branches of the circuit consists of an orifice. In accordance with the definition of the hydraulic ohm, the resistance can be determined with Equation 13.18, $R = \sqrt{(\rho/2)}(1/C_d A)$. Inspection of the expression for R for an orifice shows that the value is independent of flow. Consequently, the resistances that will be calculated in this example will invariant so the example may be solved explicitly in one step.

The system characteristics are given in Table 13.1: For the given circuit and the system characteristics given, three unknowns can be evaluated. These are the value of pump outlet pressure, p, and the flow values, Q_1 and Q_2, in the upper and lower circuit branches.

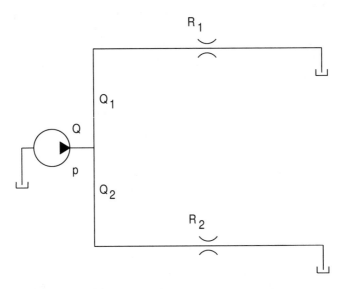

Figure 13.4: System with two flow branches.

To keep the appearance of the various expressions simpler, the area of the upper branch orifice will be evaluated explicitly:

$$A_{o1} = \frac{\pi \times (1.2E{-}3)^2}{4} = 1.131E{-}6 \text{ m}^2$$

The resistance in the upper branch is:

$$R_1 = \sqrt{\frac{\rho}{2}} \frac{1}{C_d A_o} = \sqrt{\frac{850}{2}} \frac{1}{0.6 \times 1.13E{-}6} = 30.38E{+}6$$

Perform similar calculations for the lower branch:

$$A_{o2} = \frac{\pi \times (1.0E{-}3)^2}{4} = 0.7854E{-}6 \text{ m}^2$$

and

$$R_2 = \sqrt{\frac{850}{2}} \frac{1}{0.6 \times 0.785E{-}6} = 43.75E{+}6$$

The upper and lower branch resistances are in parallel so Equation 13.24 may be used to calculate their consolidated value:

$$\frac{1}{R_T} = \frac{1}{R_1} + \frac{1}{R_2} = \frac{1}{30.38E{+}6} + \frac{1}{43.75E{+}6} = 557.4E{-}8$$

Table 13.1: Characteristics of circuit with flow independent resistances

Characteristic	Size	Units
Pump flow	0.25	L/s
Oil density	850	kg/m^3
Orifice, R_1, dia	1.2	mm
Orifice, R_1, C_d	0.6	
Orifice, R_2, dia	1.0	mm
Orifice, R_2, C_d	0.6	

Thus:

$$R_T = 17.93\text{E}+6$$

The pressure drop across the circuit can now be calculated by applying Equation 13.16, ($\sqrt{\Delta p} = RQ$), in a squared form to obtain Δp directly:

$$\Delta p = (R_T Q)^2 = (17.93\text{E}+6 \times 0.25 \times 1.0\text{E}-3)^2 = 20.09\text{E}+6 \text{ Pa}$$

Flow into an individual branch of a set of parallel legs with resistance R_r can be calculated from:

$$Q_r = \frac{\sqrt{\Delta p}}{R_r}$$

but $\sqrt{\Delta p}$ can be written:

$$\sqrt{\Delta p} = R_T Q$$

thus an alternative expression for calculating the flow into each leg of a parallel circuit is:

$$Q_r = \frac{R_T}{R_r} Q_T \qquad (13.26)$$

Using Equation 13.26 to obtain Q_1 and Q_2:

$$Q_1 = \frac{R_T}{R_1} Q_T = \frac{17.93\text{E}+8}{30.38\text{E}+6} 0.25\text{E}-3 = 0.1475\text{E}-3 \text{ m}^3/\text{s}$$

and

$$Q_2 = \frac{17.93\text{E}+8}{43.75\text{E}+6} 0.25\text{E}-3 = 0.1025\text{E}-03 \text{ m}^3/\text{s}$$

In this specific example, the three unknown pressure and flow values, p, Q_1, and Q_2, can also be obtained with use of the conventional equations:

$$Q = Q_1 + Q_2$$

$$Q_1 = C_d d_1^2 \frac{\pi}{4} \sqrt{\frac{2}{\rho}} \sqrt{p}$$

and

$$Q_2 = C_d d_2^2 \frac{\pi}{4} \sqrt{\frac{2}{\rho}} \sqrt{p}$$

These three equations may be solved simultaneously and provide the same pressure and flow values determined with hydraulic ohm and DC electrical circuit theory used above. It should noted that this alternative approach, really just manipulating the same equations in a different manner, may not be possible. First pressure across all resisting elements may not be known initially and second the value of the resistances may be flow dependent. Under such conditions the hydraulic ohm technique will generally provide an easier solution technique.

13.5.2 Example: Resistance Dependent on Flow

Consider the circuit presented in Figure 13.5. This circuit has two features

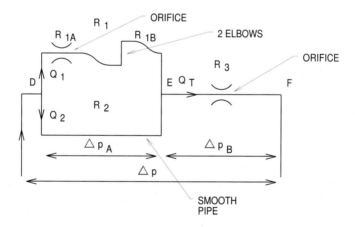

Figure 13.5: Parallel and series hydraulic circuit with flow dependent resistance.

that make it more complex to analyze than the circuit in Figure 13.4. The resistance marked R_2 on the figure is a smooth wall circular pipe. The resistance of this pipe will be given by Equation 13.13. Observe that the friction factor, f, that is, contained in the expression for R will be a function of flow, Q. Consequently this problem will need to be solved by an iterative

method that performs a step of pressure and flow estimates and then updates the resistance in the pipe using the newest information on velocity.
The other feature making the example more complex is the fact that the
two parallel legs merge to feed the series leg. In this problem, the pressure
at node E will not be known exactly until the solution is complete. The
complexity is really only superficial because the configuration allows the
four resistances present, R_{1A}, R_{1B}, R_2, and R_3 to be consolidated into one
resistance R_T. The characteristics of the circuit are presented in Table 13.2

Table 13.2: Characteristics of circuit with flow dependent resistance

Characteristic	Size	Units
Overall applied pressure, Δp	500	$lb_f/in.^2$
Oil density, ρ	0.03	$lb_m/in.^3$
Oil viscosity, μ	4.0E−6	$lb_f \cdot s/in.^2$
Orifice, R_{1A}, dia	0.16	in.
Orifice, R_{1A}, C_d	0.65	
Elbow, R_{1B}, K	1.7	
Pipe length, ℓ	720	in.
Pipe diameter, d	0.375	in.
Orifice, R_3, dia	0.22	in.
Orifice, R_3, C_d	0.61	

In keeping with the *alternation between units* philosophy of this text,
this example is being presented in inch, pound, second units. Thus the value
of density that must be used is $\rho = 0.03/386.4 = 77.6E-6\ lb_f \cdot s^2/in.^4$. The
pressures (Δp) used will be expressed in $lb_f/in.^2$ and the flows (Q) as $in.^3/s$.

As indicated earlier, this problem cannot be solved explicitly in one step
because the value of the resistance, R_2, for the smooth pipe is a function of
the Reynolds number. Initially, the flow needed to calculate the Reynolds
number is unknown. A satisfactory method of finding an initial estimate for
f is to examine Section 3.4. Most circuits employing oil operate at Reynolds
numbers less than 10000. Consequently selecting an initial f around 0.036
should be adequate. As with many iterative methods, assumptions and
compromises have to be made. In this method, the friction factor must be
available in a form that can be calculated in the algorithm. Either a table

lookup method or a functional expression can be used. In this example a functional expression was used. The laminar flow friction factor is an exact expression for any circular pipe (Equation 3.9):

$$f = \frac{64}{Re} \qquad (3.9)$$

The Blasius empirical expression was used for turbulent flow in a smooth pipe (Equation 3.10):

$$f = \frac{0.316}{Re^{0.25}} \qquad (3.10)$$

A relation (linear on the log log Moody diagram) was developed for the transition region between $Re = 2000$ and $Re = 4000$:

$$f = 0.00298 Re^{0.312}$$

The algorithm used to solve the problem had the form shown in Fig 13.6. The relative error is calculated from the following expression:

$$\text{Relative error} \; = \; \left| \frac{f_{new} - f_{old}}{f_{new}} \right|$$

Because estimation of friction factors is not very precise, it is probably not worth setting a relative error below 0.0001.

Table 13.3 shows some sequential results until convergence was reached for a relative error on f of 0.0001. The choice of variable used to detect convergence is flexible. The level of accuracy will be set by the user. It would appear that using flow, e.g., Q_T might lead to acceptable convergence more rapidly. The table shows that the method works well and converges rapidly inspite of the fact that one resistance varies with flow rate.

13.6   APPLICATION TO UNSTEADY STATE FLOW

The method is applied to steady state conditions by Esposito [1]. Application of the method to unsteady state conditions, with comparison to some laboratory results, has been published in the NCFP Proceedings [4]. Additional studies of systems under dynamic flow conditions are described in SAE papers 921686 and 932489 [5, 6]. For steady state flow conditions the parameter R is a constant for each circuit or systems element. For unsteady state flow analysis the parameter R may be a variable that must be calculated, along with all other variables, at each increment of time. The

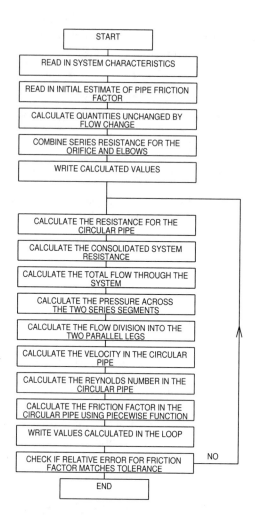

Figure 13.6: Program flowchart for flow dependent resistance example.

method avoids the need for trial and error solution of pressure and flow values in a flow network analysis, therefore, it is very useful in modeling unsteady state flow division. Minor losses can easily be included in the flow network model.

Pressurized lubrication circuits that are incorporated in machinery such

Table 13.3: Results of iteration for pipe resistance example

Variable	Step 1	Step 2	Step 3	Step 4	Step 5
f	0.0360	0.0323	0.0327	0.0326	0.0326
R_2	0.4690	0.4442	0.4469	0.4466	0.4466
R_T	0.3579	0.3537	0.3541	0.3541	0.3541
Q_1, in.^{3}/s	30.98	30.50	30.55	30.55	30.55
Q_2, in.^{3}/s	31.50	32.73	32.59	32.61	32.61
Q_T, in.^{3}/s	62.48	63.23	63.14	63.15	63.15
Δp_1, lb$_f$/in.^{2}	218.2	211.4	212.2	212.1	212.1
Δp_2, lb$_f$/in.^{2}	281.8	288.6	287.8	287.9	287.9
Re	2076.0	2157.0	2148.0	2149.0	2149.0

as transmissions and internal combustion engines make use of the same fundamental laws that apply to fluid power machinery. The available oil flow must be proportionately distributed to the branches of the lubrication gallery. An example that relates to lubrication flow distribution has been published [5].

13.6.1   Example: The Resistance Network Method Applied to Unsteady Flow

Figure 13.7 shows a fluid power system with two flow branches. The lower branch contains a compensating flow regulator valve. The valve, which is shown in Figure 13.8, has a fixed area orifice at the left end and a variable area orifice at the top. The valve is free to move in its bore due to the forces that develop from the oil pressure and spring forces. The top orifice is configured as a rectangular flow slot. Therefore, the flow area of the top orifice will vary as the valve moves in its bore. The flow slot has zero overlap and its area is a maximum when the system is not operating.

The goal of the analysis that follows is to determine suitable operating characteristics for the flow regulator valve. Therefore, the upper flow branch has an orifice that is fixed for analysis, but could be varied in size by the user to simulate a circuit workload.

Because the hydraulic resistance method will be used to study the system, an appropriate resistance term R must be established for each circuit element. Resistance values may be combined in a flow branch, or in an

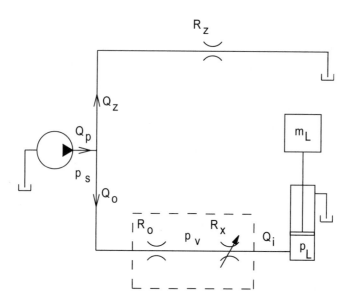

Figure 13.7: System with two flow branches.

entire system, in the same manner that is used in electric circuits [1, 7-9]. Values for flow, can then be determined with the use of the circuit equations [1, 9] that have been reviewed in Sections 13.4 and 13.5. The last part of the analysis incorporates the dynamic aspects of some of the circuit components. These components will be analyzed in the conventional fashion by applying continuity, $(dp/dt = (\beta/V)dV/dt)$ and Newton's Second Law, $(F = ma)$. The solution to the problem will require that the differential equations with displacement and pressure as the state variables be solved iteratively. As the solution moves forward in time, the values of resistance and flow will be updated automatically. The variables that are used to describe the operation of the system are presented in Table 13.4.

We shall start the analysis by developing the resistance terms for the flow regulator valve and the actuator. Expressions must be written for the resistance terms R_o, R_x, and R_L. The first two are of the orifice flow type and may be written as:

$$R_o = \sqrt{\frac{\rho}{2}} \frac{1}{C_d A_o}$$

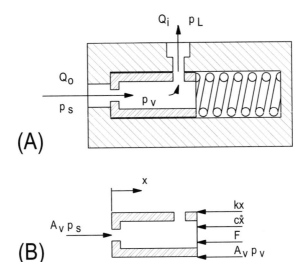

Figure 13.8: (A) Compensating flow regulator valve and (B) Forces on the valve spool.

and:

$$R_x = \sqrt{\frac{\rho}{2}} \frac{1}{C_d w(\ell_0 - x)}$$

As explained in Section 13.4, the resistance for a actuator can be written as:

$$R_L = \frac{\sqrt{p_L}}{A_p \dot{y}}$$

The series resistance terms in the lower flow branch, as shown on Figure 13.7, may be combined into a single resistance, R_{oxL}:

$$R_{oxL} = \sqrt{R_o^2 + R_x^2 + R_L^2}$$

The resistance in the upper branch, R_z, is simply an orifice so:

$$R_z = \sqrt{\frac{\rho}{2}} \frac{1}{C_d A_z}$$

Total parallel resistance R_T, the total system resistance seen by the pump, may be expressed as:

$$R_T = \frac{R_z R_{oxl}}{R_z + R_{oxl}}$$

Table 13.4: Characteristics for flow division analysis

Characteristic	Description	Units
p_s, p_v, p_L	System pressure values	N/m^2
Q_p, Q_i, Q_o, Q_z	System flow values	m^3/s
A_z	Variable load orifice area	m^2
A_v	Valve end area	m^2
m_V	valve mass	kg
A_o	Fixed valve orifice	m^2
w	Valve flow gradient	m^2/m
k	Valve spring rate	N/m
F	Valve spring preload	N
c	Damping coefficient	N.s/m
C_d	Flow coefficient	
ℓ_0	Length of flow slot	m
V_c	Oil volume inside valve	m^3
ρ	Oil mass density	kg/m^3
g	Acceleration of gravity	m/s^2
β_e	oil bulk modulus	N/m^2
A_p	Cylinder piston area	m^2
d	Cylinder piston diameter	m
ℓ_{sl}	Piston seal width	m
μ_{st}	Coefficient of static friction (seal)	
V_L	Cylinder oil volume	m^3
m_L	Cylinder load mass	kg
R_o, R_z, R_x, R_L	Element flow resistance	Pa$^{1/2}$s/m^3
R_T, R_{oxl}	Combined resistance	Pa$^{1/2}$s/m^3
x	Valve motion	m
y	Cylinder piston motion	m

Most of the pressures in the circuit will be state variables derived from $dp/dt = (\beta_e/V)dV/dt$ expressions. An exception is the system pressure p_s. In this problem, the driving function is the pump flow so the system pressure will need to be upgraded at each step of the the iterative solution. The hydraulic ohm method will be used. Values for pressure at the pump outlet p_s, are given by:

$$p_s = R_T^2 Q_p^2(t)$$

The flow in the upper system branch Q_z may then be determined as:

$$Q_z = \frac{R_T}{R_z} Q_p(t) \tag{13.27}$$

There would appear to be two methods of determining Q_o. Both approaches involve using a resistance method:

$$Q_o = \frac{R_T}{R_o} Q_p(t)$$

or Q_o could be found from:

$$Q_o = \frac{\sqrt{p_s - p_v}}{R_o} \tag{13.28}$$

Because the major part of the solution involves solving for state variables using a differential equation solver, it is usually desirable to employ the state variables as much as possible. It will be shown shortly that p_v is a state variable, so Equation 13.28 is the preferred equation. Another reason to consider in choosing Equation 13.28 is that this formulation only involves one resistance, R_o, and not the three, R_o, R_x, and R_L, that are needed to calculate R_T. The remaining flow that must be determined before the dynamic equations can be formulated is the flow from the valve into the load, Q_i. This flow is determined from:

$$Q_i = \frac{\sqrt{p_v - p_L}}{R_x} \tag{13.29}$$

Now that the flows in and out of the various components have been established using the hydraulic ohm method, the dynamic aspects of the system can be evaluated. Values for the cylinder pressure p_L, and the pressure p_v inside the control valve, can be established with use of the continuity of flow principle. Flow continuity, as it applies to the cylinder and the flow control valve, includes flow into and out of the confined volumes V_L and V_c. Also, the influence of fluid compressibility and the effect of moving parts must be included. Flow continuity applied to volume V_c may be written as:

$$Q_o = Q_i - A_v \dot{x} + \frac{V_c}{\beta_e} \dot{p}_v$$

Where flows Q_o and Q_i have been determined in Equations 13.28 and 13.29. The volume V_c is essentially constant since motion x for the valve is small. The equation for continuity may be rearranged as follows:

$$\frac{d}{dt}p_v = \dot{p}_v = \frac{\beta_e}{V_c}(Q_o - Q_i + A_v\dot{x}) \tag{13.30}$$

The notation dp_v/dt has been used to show that p_v is a state variable, i.e., one variable in a set of ordinary differential equations that are being developed to complete solution of the problem.

Flow continuity applied to the volume V_L may be written as:

$$Q_i = A_p\dot{y} + \frac{V_L}{\beta_e}\dot{p}_L$$

The cylinder oil volume V_L is a function of the cylinder piston motion y and may be expressed as:

$$V_L = A_p y$$

This equation may be rearranged as follows to provide another equation in the set.

$$\frac{d}{dt}p_L = \dot{p}_L = \frac{\beta_e}{A_p y}(Q_i - A_p\dot{y}) \tag{13.31}$$

A schematic for the flow regulator valve is shown in Figure 13.8 (panel A). The flow control element, afterwards just called a valve, must be free to move. The equation of motion for the valve can be established from the forces shown on the valve free body diagram shown in Figure 13.8 (panel B). From Newton's Second Law:

$$A_v p_s - A_v p_v - kx - F - c\dot{x} = m_v\ddot{x}$$

This equation can be rearranged so that it can be integrated to find the valve velocity and valve position x:

$$\frac{d}{dt}\dot{x} = \ddot{x} = (A_v p_s - A_v p_v - kx - F - c\dot{x})/m_v \tag{13.32}$$

Incidentally a further differential equation must be written:

$$\frac{d}{dt}x = \dot{x} \tag{13.33}$$

because most differential equation solvers only work with sets of first order equations.

A reasonable value for the damping coefficient c has been established from laboratory test results. This value applies to spring loaded valves operating in close fitting bores. For this purpose the damping coefficient can be expressed as a fraction of the theoretical critical damping as:

$$c = 0.5(2\sqrt{km_v}) \tag{13.34}$$

For most valves of this type, the proper fraction shown here as 0.5, will vary from 0.5 to 0.7.

The equation of motion for the cylinder piston and load can be established from the free body diagram shown in Figure 13.9:

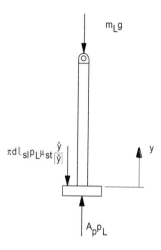

Figure 13.9: Forces on cylinder piston and load controlled by a flow regulator valve.

$$A_p p_L - m_L g - \pi d\ell_{sl} p_L \mu_{st} \frac{\dot{y}}{|\dot{y}|} = m_L \ddot{y}$$

To determine load piston velocity $\dot{y}$ and position y the equation may be arranged as:

$$\frac{d}{dt}\dot{y} = \ddot{y} = \left(A_p p_L - m_L g - \pi d\ell_{sl} p_L \mu_{st} \frac{\dot{y}}{|\dot{y}|} \right) / m_L \tag{13.35}$$

As before, add the following equation to the set:

$$\frac{d}{dt}y = \dot{y} \tag{13.36}$$

A value for the damping coefficient is difficult to determine for a cylinder piston. It is well known from laboratory work, however, that a friction force exists at the piston and rod seals. In the above equation the friction has been modeled as a single value that incorporates piston diameter d, seal width ℓ_{sl}, load pressure p_L, and seal Coulomb friction μ_{st}. The friction force always opposes motion of the piston. Therefore, the direction of the friction force must be corrected with the ratio of piston velocity over absolute velocity $\dot{y}/|\dot{y}|$.

The example that we have examined requires a set of steps for its solution. There are many packages that can be used for solution and each may have different methods of interaction between the user and the package. All, however, must follow the same internal sequence. Sufficient initial values of flow and pressure must be submitted in addition to the system, i.e., component, characteristics. The solution formulation is dynamic, that is, a set of simultaneous, ordinary differential equations with pressures, displacements, and velocities as state variables is solved numerically. The user must ensure that the coefficients in these state variable equations that depend on varying resistance values are updated during the iteration process. A generic flowchart is presented in Figure 13.10 showing the calculation sequence that is necessary.

13.6.2 Example Results and Discussion

The mathematical model, which has been established for the system, was solved with parameter values as shown in Table 13.5. Input flow $Q_p(t)$ to the system was started at time equal to zero at $0.2\text{E}{-}3 \text{ m}^3/\text{s}$. At time equal to 0.1 s the flow was increased linearly to a value of $0.2133\text{E}{-}3 \text{ m}^3/\text{s}$ at 2 s.

It should be noted that selection of input values requires some care and judgement. The performance of the actuator, i.e., the extension and rate of extension, were fixed by machine operating requirements. The static pressure required to raise the load was easily calculated. These values were used to estimate the fixed and variable orifices in the flow regulator valve. First estimates of the preload and the spring rate could also be obtained from static considerations. It had been decided that the flows in the two branches of the circuit should be approximately equal. After obtaining the preliminary estimates, the system was simulated and certain values adjusted until the system could be integrated over the desired time interval. It was found that the simulation was quite sensitive to the spring rate value. The value of $k = 2700 \text{ N/m}$ was the lowest value that allowed operation without integration failure or the valve closing prematurely. Results are shown in Figures 13.11 to 13.14 for the 2 s time interval. The plotted results show the valve motion x, actuator motion y, pressure and flow at various points in the circuit. It will be observed that the system performs quite well

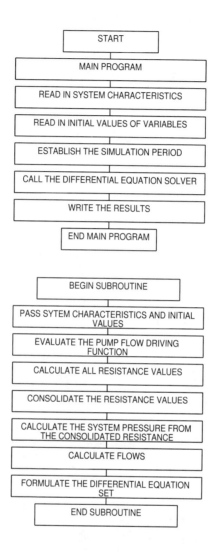

Figure 13.10: Program flowchart for unsteady flow example.

even at this lowest spring rate. The plot of the load pressure p_L vs. time, (Figure 13.13) is essentially constant even when the pump flow Q_p and system pressure p_s are increasing.

Variations in the valve spring rate k, and its preload F, provide the easiest approach to adjusting system performance. The spring rate of k

Table 13.5: Values and appropriate units applied for the example

Characteristic	Value	Units
Variable load orifice area, A_z	0.903E−6	m^2
Valve end area, A_v	0.129E−3	m^2
Valve mass, m_v	0.023	kg
Fixed valve orifice, A_o	6.45E−6	m^2
Valve flow gradient, w	2.5E−3	m^2/m
Valve spring rate, k	2700	N/m
Valve spring preload, F	32	N
Orifice flow coefficient, C_d	0.6	
Length of flow slot, ℓ_0	4.0E−3	m
Oil volume in valve, V_c	33E−6	m^3
Oil mass density, ρ	855	kg/m^3
Acceleration of gravity, g	9.8	m/s^2
Oil bulk modulus, β_e	1.172E+9	N/m^2
Cylinder piston diameter, d	48.3E−3	m
Piston seal width, ℓ_{sl}	2.5E−3	m
Coefficient of static friction (seal), μ_{st}	0.05	
Cylinder load mass, m_L	2140	kg

$= 2700$ N/m allowed little margin for unexpected changes in actuator performance because the valve spool excursion was nearly 85% of the allowed maximum. This opening could be reduced by increasing the valve spring rate. Figures 13.15 to 13.18 show the system performance with the spring rate increased to $k = 5000$ N/m. As expected, the valve opens less at the end of the simulation. The actuator extension is slightly greater because the orifice area, A_x, is larger. A benefit of the stiffer spring is that the system pressure is slightly reduced. The pressure in the actuator remains constant and unchanged. Minor disadvantages of using the stiffer spring are that the actuator flow rises slightly towards the end of the simulation and the flow to the upper branch is slightly less. This reduction is a consequence of the lower system pressure.

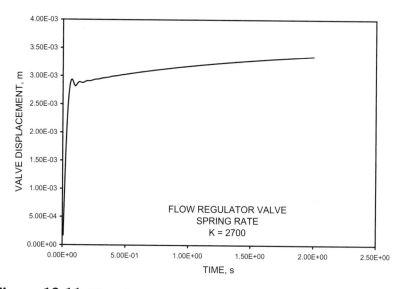

Figure 13.11: Flow division example, valve spool motion, x, vs. time, t.

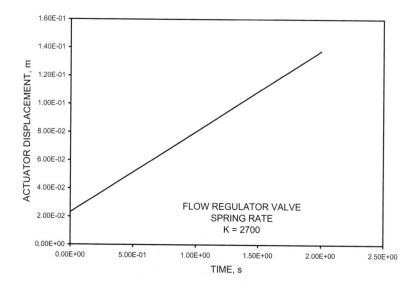

Figure 13.12: Flow division example actuator motion, y, vs. time, t.

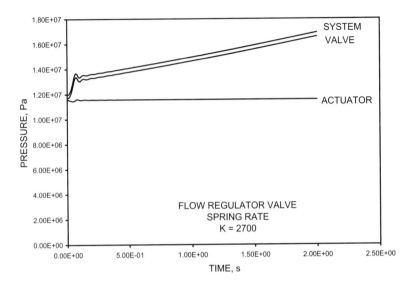

Figure 13.13: Flow division example showing system, p_s, valve, p_v, and actuator, p_L, pressures vs. time, t.

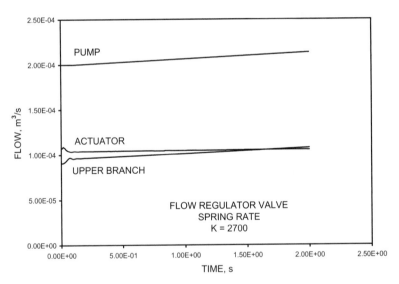

Figure 13.14: Flow division example showing pump, Q_p, actuator, Q_i, and upper branch, Q_z, flows vs. time, t.

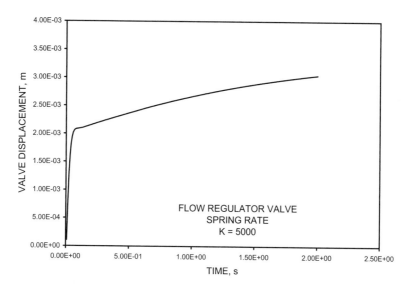

Figure 13.15: Flow division example valve spool motion, x, vs. time, t.

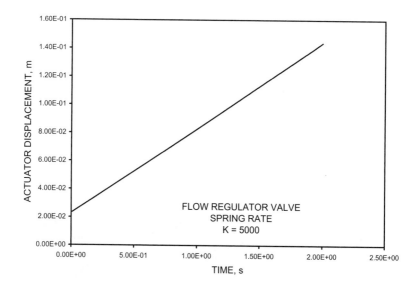

Figure 13.16: Flow division example actuator motion, y, vs. time, t.

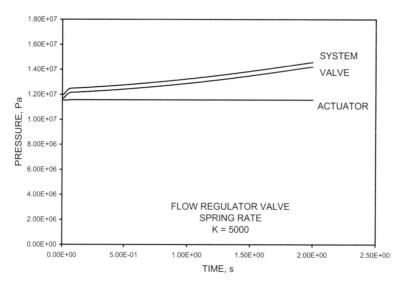

Figure 13.17: Flow division example showing system, p_s, valve, p_v, and actuator, p_L, pressures vs. time, t.

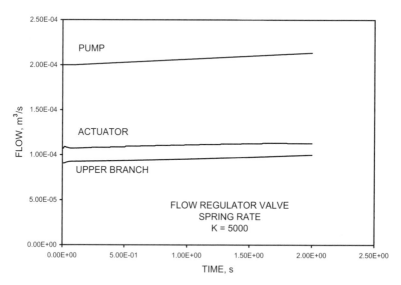

Figure 13.18: Flow division example showing pump, Q_p, actuator, Q_i, and upper branch, Q_z, flows vs. time, t.

13.7 CONCLUSIONS

The type of mathematical model presented is very useful in sizing parameters to provide the desired operating characteristics of a flow regulator valve. Many parameter variations could be applied to the model. The major purpose, presented here, is to define the mathematical model and illustrate its usefulness in component design. Many methods are available to solve the equations, a symbolic computer software program was used for the solution presented here [10].

Values for the fluid flow variables Q_z, and Q_i, and pressure p_s at the pump outlet were determined with use of the hydraulic ohm principle. These values may also be determined with a simultaneous solution of the appropriate equations that contain these parameters. Numerical values determined with the simultaneous solution are identical to those calculated with use of the ohm principle.

Because computer solution involves iteration, that is usually small time steps are made towards a final solution, the hydraulic resistance terms can be updated at each iteration step using the current pressure and flow values. It has been found in practice that the small steps generally required to solve the dynamic equations in dx/dt, $d\dot{x}/dt$, and dp/dt allow adequate estimation of resistance values without needing *intrastep* iteration for the resistance terms [4-6].

Numerous combinations of equations are possible. Solution of the equations with the hydraulic ohm method, however, consistently provides the same results as solution by simultaneous methods. Because both methods make use of the same principles from fluid mechanics, consistent answers may be expected.

Values for the flow variables Q_z, and Q_o can be determined with the simultaneous solution of the four equations that follow:

$$Q_o = C_d A_o \sqrt{\frac{2}{\rho}} \sqrt{p_s - p_v}$$

$$Q_i = C_d w (h - x) \sqrt{\frac{2}{\rho}} \sqrt{p_v - p_L}$$

$$Q_z = C_d A z \sqrt{\frac{2}{\rho}} \sqrt{p_s}$$

$$Q_p = Q_o + Q_z$$

These flow values can then be used to calculate the coefficients in the state variable equations in pressure, displacement, and velocity (Equations 13.30, 13.31, 13.32, 13.33, 13.35, and 13.36).

It may be observed that simultaneous solution of the equations in Q_o, Q_i, Q_z, and Q_p is easy in this example because the resistance R_z is invariant. More complex branches with different components might lead to simultaneous equations in various Q values that would be more difficult to solve. Under such conditions, the hydraulic ohm method may be much easier to apply.

PROBLEMS

13.1 Flow, Q, from the pump divides through the two branches as shown in the figure.

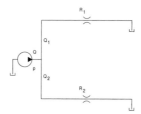

Write a set of equations, based on conventional fluid flow theory, that can be used to solve for pressure, p, and the flow values, Q_1 and Q_2. Neglect line friction loss.

Characteristics of flow division with two fixed orifices

Characteristic	Size	Units
Upper branch orifice diameter, d_1	0.048	in.
Lower branch orifice diameter, d_2	0.040	in.
Orifice flow coefficient (both), C_d	0.6	
Oil mass density, ρ	0.78E−4	$\mathrm{lb_m \cdot s^2/in.^4}$

13.2 Use the equations determined in Problem 13.1 to solve for numerical values of pressure, p, and flow values, Q_1 and Q_2, when pump flow, Q, is equal to 4.00 gpm.

13.3 Use the information given for exercises Problems 13.1 and 13.2 to obtain numerical values for pressure, p, and flow values, Q_1 and Q_2, with use of the hydraulic ohm method.

13.4 Leakage input flow, $Q = Q_1 + Q_2$, passes through the spool valve shown in the figure.

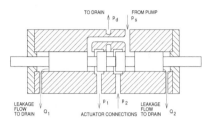

The oil flows past the two spool lands in annular passages as, Q_1 and Q_2. Write a set of equations, based on conventional fluid flow theory, that can be used to solve for pressure, p_s, and the flow values, Q_1 and Q_2.

Characteristics of flow division in a spool valve

Characteristic	Size	Units
Spool diameter, d_{sp}	20.0	mm
Land length, ℓ_1	93.0	mm
Land length, ℓ_2	73.0	mm
Spool radial clearance (centered)	0.025	mm

13.5 Use the equations determined in Problem 13.4 to solve for numerical values of pressure, p_s, and flow values, Q_1 and Q_2 when the leakage input flow, $Q = Q_1 + Q_2$, is equal to 0.01 L/s.

13.6 Use the information given for exercises Problems 13.4 and 13.5 to obtain numerical values for pressure, p_s, and leakage flow values, Q_1 and Q_2, with use of the hydraulic ohm method.

13.7 Model the flow regulator valve discussed in this chapter with any available simulation method and compare the results to the published results.

13.8 A manufacturing machine includes a fluid power system with a motor and cylinder operating in parallel.

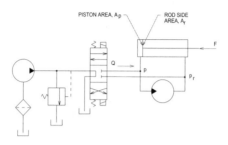

Determine the time, t, in seconds that is required before the cylinder piston begins to move for the conditions given in the table. The motor torque, T, increases with time, t, seconds. Determine the pressure, p, that has developed in the system for the time, t, calculated for the previous part of the question. Determine the initial velocity, $\dot{x}$, for the cylinder piston for the given pump flow, Q, and with motor speed, n, equal to 500 rpm when the piston begins to move.

Characteristics of a manufacturing machine system

Characteristic	Size	Units
Cylinder piston area, A_p	1800	mm^2
Cylinder rod side area, A_r	1290	mm^2
Cylinder load force, F	4500	N
Motor displacement, D_m	72.0	mL/rev
Motor discharge pressure, p_r	695	kPa
Motor output torque, T	$10.0\,t^{1.4}$	N $\cdot$ m
Pump flow rate $0 < t < 2.1$, Q	0.496 t	L/s
Pump flow rate $2.1 < t$, Q	1.0	L/s

REFERENCES

1. Esposito, A., 1969, "A Simplified Method for Analyzing Circuits by Analogy", Machine Design, October, pp. 173-177.

2. The Lee Company, 1987, *Lee Technical Hydraulic Handbook*, The Lee Company, Westbrook, CN, pp. 367-397.

3. Del Toro, V., 1972, *Principles of Electrical Engineering*, 2nd ed. Prentice-Hall, Inc., Englewood Cliffs, NJ.

4. Gassman, M. P., 1978, "Prediction of Pressure and Flow Values for a Hydraulic System", *Proceedings of the 34th National conference on Fluid Power*, NFPA, Philadelphia, PA, pp. 377-382.

5. Gassman, M. P., 1992, "Fluid Power System Flow Distribution and Components Analysis", SAE Paper 921686, SAE, Milwaukee, WI.

6. Gassman, M. P., 1993, "Use of the Hydraulic Ohm to Determine Flow Distribution", SAE Paper 932489, SAE, Milwaukee, WI.

7. L. Dodge, 1968, "How to Compute and Combine Fluid Flow Resistances in Components, Part 1", Hydraulics and Pneumatics, September, p. 118.

8. L. Dodge, 1968, "How to Compute and Combine Fluid Flow Resistances in Components, Part 2", Hydraulics and Pneumatics, November, p. 98.

9. Martin, H., 1995, *The Design of Hydraulic Components and Systems*, Ellis Horwood, New York, NY, pp. 204-225.

10. MathSoft, Inc., 2001, "Mathcad, User's Guide with Reference Manual", Cambridge, MA.

14

NOISE CONTROL

14.1 INTRODUCTION

Noise in fluid power systems is generally caused by pressure waves in the fluid stream. The pumps in general use are positive displacement types and employ pistons, vanes, or gear teeth to move the fluid. Since the fluid moves in sequential packets, pressure waves are set up in the fluid stream. These pulses of fluid are a cause of noise and system instability [1, 2].

Many of the valves used to control hydraulic systems are constrained with steel coil springs. The valves are free to move within the valve housing and are biased on one or both ends with springs. The valve and spring effectively act as a spring-mass system. The valve may then oscillate at its own natural frequency and is also subject to the forcing function provided by the pressure waves in the fluid stream. Pressure waves in hydraulic systems can cause control valves to become unstable during operation and also contribute to vibration and noise. Therefore, it becomes desirable to filter out or at least reduce the magnitude of the pulses, in order to optimize the performance of fluid power systems and their controls. Reduction in pressure wave amplitude also reduces wear and damage to system parts.

The fluid pump is usually the primary source of pressure pulsations. These waves travel throughout the fluid system. Therefore, it becomes advantageous to reduce the amplitude of the pressure waves as close to the source as possible [3, 4].

A method of reducing the pressure waves by use of a carefully selected volume in the flow line is described. A successful mathematical means of sizing the desired attenuator volume is outlined. The mathematical model can be solved with a variety of computer simulation programs. A properly sized attenuator volume is useful over a range of frequencies and pressure amplitudes, which makes it desirable for many systems. The method was developed in the agricultural tractor manufacturing industry and has proved to be successful in application.

14.2 DISCUSSION OF METHOD

An inherent advantage of fluid power systems is the ability to transmit energy through flow lines that can be configured to any desired machine. Unstable operation of a system, however, can result as pressure disturbances are produced. Most of the fluid pumps in general use cause pulsations in flow lines. Amplification of these pulses sometimes occurs in lines, causing system oscillation and noise.

Many systems have feedback controls or carefully designed valves, all of which can be disturbed by pressure waves. Stability can be further aggravated by flow in lines causing time delay that produces phase shift.

One of the most effective means of reducing the amplitude of fluid pressure pulsations is to add a fluid volume to a flow line [5-7]. Pressure and flow variations are part of the same fluid phenomenon and are clearly associated. As a varying flow passes through the volume, the traveling fluid compresses to store additional fluid during above average flow periods and then releases the fluid during lower than average flow periods. This effect causes attenuation of the volume outlet flow variations and the associated pressure pulsations. This method of attenuation is relatively inexpensive and easy to apply to a great variety of fluid flow systems.

An understanding of fluid line dynamics can be gained by considering the correlation to reciprocating engine intake and exhaust systems. Properly tuned systems will help scavenge the combustion chamber of exhaust gases, increase the volumetric efficiency, and reduce noise. Similar benefits can be gained for a fluid power system with the proper tuning of flow lines [8].

A method for tuning fluid power systems is presented in the following sections. The approach has room for improvement, but can bring considerable improvement to many systems as presented.

The mathematical approach presented can be used to select an attenuator volume that will be most beneficial at the pump's primary, or rated, operating speed. Information is given in Section 14.3.4 on a practical application that relates to a satisfactory use of a tuned fluid volume.

14.3   MATHEMATICAL MODEL

The suite of equations arising from the continuity equation, Navier-Stokes equations, the energy equation, and the equation of state for a fluid would have to be solved to provide an exact description of fluid line dynamics. Solution of these equations is difficult. The resulting solution, providing it could be obtained, would be too complex for ordinary engineering use. A great variety of work has been done to analyze fluid line dynamics and present suitable mathematical approaches [1, 9-11]. Many of the resulting systems of equations are complex. Fortunately, measurable improvement can be gained in fluid line systems with the somewhat simplified approach described here.

If the relevant assumptions are simplified, several different types of models can be developed. In order to provide timely improvements for typical fluid power systems, it becomes desirable to develop the simplest set of equations that can provide a reasonable description of the performance of the machine under consideration.

Good results can be obtained in the study, and subsequent improvement, of a system of fluid lines by use of the relatively simple lumped parameter model. Fluid compressibility, fluid column inertia, and resistance to flow must be considered to develop a reasonable mathematical model to represent an attenuator and its attached lines. Development of the equations requires that these three factors act independently of each other. Each of the three effects must be integrated over the length of the line and lumped into discrete parts. The resulting model is an approximation that provides reasonably good results for practical applications.

If comparatively long line lengths occur in a system, the lumped parameter model allows easy experimentation with two or more tuned volumes in a given line. Because the mathematical model makes use of an electric-hydraulic analogy, additional volumes in a line simply add more loops to the circuit model. The model allows sizing and locating one or more volumes in a long line to provide the desired pressure wave amplitude reduction.

A distinct advantage of the lumped parameter approach is that the mathematical model is made up of ordinary differential equations. The resulting equations can be easily solved. Also, transient and frequency responses are readily obtained [11].

14.3.1   Derivation of Fluid Analogies to Resistance, Inductance, and Capacitance

Electrical and fluid analogies: As indicated in Chapter 13, there is an analogy between electrical and fluid circuits. In this chapter, the resistance elements are circular pipes and the flow through these pipes will be considered to be laminar. The advantage of limiting flow to the laminar regime is that the fluid resistance, inductance, and capacitance are almost exactly parallel to their electrical counterparts.

Please note, however, that the definition of hydraulic resistance used in this chapter differs from that used in Chapter 13. Resistance in Chapter 13 was related to turbulent flow where it was more convenient to use a relationship of the form $\sqrt{\Delta p} = RQ$. In this chapter, the analogy with electrical circuits exposed to sinusoidal potentials is being used and the analogy is closer when the flow is laminar where $\Delta p = RQ$. There is no real conflict between the two definitions. It was explained in Chapter 13 that laminar flow components could be included in the calculations for flow division, but the equivalent resistance value based on $\sqrt{\Delta p} = Rq$ would be recalculated for the prevailing values at each iteration step.

In an electrical system, charges move when subjected to electrical potential. The flow of charges is electrical current. A device that converts electrical energy into heat energy is a resistor. A device that can store electrical charges and release them as the potential across the device changes is a capacitor. A device that can store and release electrical energy in the form of a magnetic field is an inductor.

In a fluid flow system, the analogy of electrical potential is pressure difference and the analogy of charge is volume. Thus flow rate is the analogy of electrical current. Devices like orifices and pipes convert pressure energy into heat. As explained in Chapter 13, orifices and pipes can be treated as resistors in much the same way as electrical resistors. Volume can be accumulated and released because fluids and their containing structures are elastic. As explained in Chapter 2, the elastic characteristic for fluids is bulk modulus. Thus any volume of fluid and its container may be treated as a fluid capacitance. The other means in which mechanical, as opposed to heat energy, can be stored and exchanged in a fluid system is essentially by means of converting a change in pressure to a change in kinetic energy and the reverse. In fact, the expression for fluid inductance is derived from Newton's Second Law, $F = ma$, because this matches the electrical definition of inductance.

Resistance and inductance for laminar flow in a pipe:
Consider the parallel expressions for electrical and fluid flow presented in

Table 14.1. The fluid flow expressions for resistance, capacitance and inductance are derived directly from the electrical definitions. As mentioned earlier, only resistance related to the laminar flow in circular pipes is presented. This has been done because the electrical and fluid analogies then become exactly parallel and many tools used to analyze electrical circuits can be used directly for their fluid analogies. Although turbulent flow devices will exhibit resistance, capacitance, and inductance, finding the performance of such devices will require computer simulation. In a sense, the expressions provided in Table 14.1 are the counterparts of the linearization of systems discussed in Chapters 5 and 6. Components for noise reduction can be sized initially using the laminar expressions.

The derivation of fluid inductance in Table 14.1 made the assumption that the fluid would move as a solid rod with the same velocity across the section of the pipe. The velocity distribution across a circular pipe for laminar flow, however, is parabolic because of the viscous effects. As with many electrical devices, it is not possible to manufacture an ideal fluid inductance. Whenever a real fluid with viscosity moves in a pipe, resistance, capacitance, and inductance effects, *all* occur simultaneously in the same same pipe. Fortunately the capacitance effect is not affected by the resistance and inductance effects and may be ignored in the derivation of inductance.

Consider the laminar flow in a circular pipe shown in Figure 14.1. Assume that the pressure difference across a length ℓ of the pipe is varying with time, so partial derivatives must be used. The velocity gradient at some radius r can be written as $\partial v/\partial r$. Now use the definition of absolute viscosity in Equation 2.1 to write the shear stress on the outer surface of the fluid rod of radius r:

$$\tau_r = \mu \frac{\partial v}{\partial r} \tag{14.1}$$

and use Equation 14.1 to develop an expression for the total viscous force on the elemental annulus shown in Figure 14.1:

$$F_{visc} = 2\pi(r + \Delta r)\ell\mu\left(\frac{\partial v}{\partial r} + \frac{\partial^2 v}{\partial r^2}\Delta r\right) - 2\pi r\ell\mu\frac{\partial v}{\partial r}$$

Ignoring $(\Delta r)^2$ terms, the expression for F_{visc} may be written:

$$F_{visc} = 2\pi\ell\mu\Delta r\left(\frac{\partial^2 v}{\partial r^2}r + \frac{\partial v}{\partial r}\right) \tag{14.2}$$

Note that F_{visc} is positive in the direction shown on Figure 14.1. The force is positive in that direction because the gradient $\partial v/\partial r$ is assumed to be positive in the direction of increasing radius r because this would be consistent with the conventions of calculus.

Table 14.1: Analogy between electrical and fluid resistance, inductance, and capacitance

Electrical	Fluid Flow

Potential and Flow Variables

Potential, V	Pressure, Δp
Current, i	Flow rate, Q

Resistance

$V = Ri$	$\Delta p = \dfrac{128\mu\ell}{\pi d^4} Q$ (laminar)
	$\Delta p = \dfrac{8\mu\ell}{\pi r_o^4} Q$
R	$R = \dfrac{8\mu\ell}{\pi r_o^4}$

Capacitance

$\dfrac{dV}{dt} = \dfrac{1}{C} i$	$\dfrac{dp}{dt} = \dfrac{\beta}{V}\dfrac{dV}{dt} = \dfrac{\beta}{V} Q$
C	$C = \dfrac{V}{\beta} = \dfrac{\ell\pi r_o^2}{\beta}$

Inductance

$V = L\dfrac{di}{dt}$	$F = ma$
	$(\Delta p \pi r_o^2) = (\rho r_o^2 \ell)\dfrac{dv}{dt}$
	$(\Delta p) = (\rho \ell)\dfrac{dQ}{A dt}$
	$\Delta p = \dfrac{\rho \ell}{\pi r_o^2}\dfrac{dQ}{dt}$
L	$L = \dfrac{\rho \ell}{\pi r_o^2}$ (plug flow)

Now apply Newton's Second Law to the elemental annular tube:

$$F_{visc} + \Delta p(2\pi r \Delta r) = (\rho 2\pi r \Delta r \ell)\frac{\partial v}{\partial t}$$

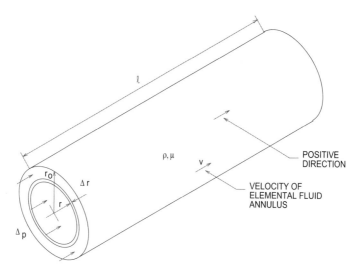

Figure 14.1: Fluid flowing in circular pipe, laminar flow regime.

$$2\pi\ell\mu\Delta r\left(\frac{\partial^2 v}{\partial r^2}r + \frac{\partial v}{\partial r}\right) + \Delta p(2\pi r\Delta r) = (\rho 2\pi r\Delta r\ell)\frac{\partial v}{\partial t}$$

$$\ell\mu\left(\frac{\partial^2 v}{\partial r^2} + \frac{\partial v}{\partial r}\frac{1}{r}\right) + \Delta p = \rho\ell\frac{\partial v}{\partial t} \qquad (14.3)$$

We shall shortly show that F_{visc} is negative because $\partial v/\partial r$ is negative for a viscous fluid flowing in a circular pipe. Thus, as might have been expected, Equation 14.3 shows that for some Δp acting on a fluid in a circular pipe, only a certain amount of that Δp is available to cause acceleration $\partial v/\partial t$.

Before using Equation 14.3 to find a more explicit relationship among the various variables, we shall return to an electrical analogy. This will be a useful guide for developing the relationship for the fluid variables. Consider the circuit shown in Figure 14.2 and use the potential and relationships given in Table 14.1.

$$V(t) = Ri\,\sin\omega t + Li\,(\omega\cos\omega t)$$

$$= i\,(R\sin\omega t + L\omega\cos\omega t) \qquad (14.4)$$

Now consider a uniform length of circular pipe containing a typical hydraulic oil that will possess viscosity. It may not be obvious if the fluid

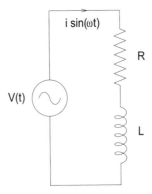

Figure 14.2: Resistance and inductance circuit excited by sinusoidal input.

model has the resistance in series or parallel with the inductance. The following reasoning should satisfy the reader that the resistance is in series. If the resistance were in parallel with the inductance, then a very low value of resistance would effectively cut out the inductance effect. If the pipe is examined, it is easily seen that reducing the resistance would not limit the inductance. Thus the series model examined for the electrical situation is also suitable for the fluid.

Suppose the oil in the pipe is subjected to a sinusoidal pressure variation. We shall assume that the pressure causes a velocity distribution that obeys the partial differential equation presented as Equation 14.3. It will be assumed that the velocity has a parabolic distribution across the pipe cross section and that the variation with time is sinusoidal. Thus consider a trial solution:

$$v = v_{ripple} \frac{r_o^2 - r^2}{r_o^2} \sin \omega t \qquad (14.5)$$

Consider a flow rate, Q, that varies with time and with radial position across the cross section. We can write:

$$Q(t) = \int_0^{r_o} \frac{\partial Q}{\partial r} dr \qquad (14.6)$$

Using Equation 14.5 shows that:

$$\frac{\partial Q}{\partial r} = 2\pi \left(v_{ripple} \frac{r_o^2 - r^2}{r_o^2} \sin \omega t \right) r \qquad (14.7)$$

Table 14.1 shows that fluid inductance is a function of Δp and dQ/dt.

Equation 14.6 can be partially differentiated with respect to t to yield dQ/dt:

$$
\begin{aligned}
dQ/dt &= \frac{\partial}{\partial t}\int_0^{r_o}\frac{\partial Q}{\partial r}dr = \int_0^{r_o}\frac{\partial}{\partial t}\frac{\partial Q}{\partial r}dr \\
&= \int_0^{r_o} 2\pi\left(v_{ripple}\frac{r_o^2 - r^2}{r_o^2}\,\omega\cos\omega t\right)r\,dr \\
&= \int_0^{r_o} 2\pi\frac{\partial v}{\partial t}r\,dr \qquad\qquad (14.8)
\end{aligned}
$$

Now combine Equations 14.3 and 14.8 to yield an equation that covers flow across the whole pipe:

$$
\int_0^{r_o} 2\pi\left(\mu\ell\left(\frac{\partial^2 v}{\partial r^2} + \frac{\partial v}{\partial r}\frac{1}{r}\right) + \Delta p\right)r\,dr = \rho\ell\int_0^{r_o}2\pi\frac{\partial v}{\partial t}r\,dr \qquad (14.9)
$$

Values of $\partial v/\partial t$ and $\partial^2 v/\partial r^2$ may be derived from Equation 14.5:

$$
\frac{\partial v}{\partial r} = v_{ripple}\frac{-2r}{r_o^2}\sin\omega t \qquad\qquad \frac{\partial^2 v}{\partial r^2} = v_{ripple}\frac{-2}{r_o^2}\sin\omega t
$$

Making these partial derivative substitutions into Equation 14.9:

$$
\int_0^{r_o}2\pi\left(\frac{-4\mu\ell}{r_o^2}v_{ripple}\sin\omega t + \Delta p\right)r\,dr =
$$

$$
\rho\ell\int_0^{r_o}2\pi\left(v_{ripple}\left(\frac{r_o^2 - r^2}{r_o^2}\right)\omega\cos\omega t\right)r\,dr
$$

$$
\frac{-4\mu\ell}{r_o^2}\left(\frac{r_o^2}{2}\right)v_{ripple}\sin\omega t + \Delta p\left(\frac{r_o^2}{2}\right) = \rho\ell v_{ripple}\left(\frac{r_o^4}{4r_o^2}\right)\omega\cos\omega t \quad (14.10)
$$

Reorganize Equation 14.10 in the form of the electrical expression for the series circuit (Equation 14.4):

$$
\Delta p = \frac{4\mu\ell}{r_o^2}v_{ripple}\sin\omega t + \frac{r_o^2}{2r_o^2}v_{ripple}\omega\cos\omega t \qquad (14.11)
$$

Noting that the relation between the flow and the maximum velocity in the pipe axis is:

$$
Q_{ripple}\sin\omega t = \frac{v_{ripple}}{2}\pi r_o^2\sin\omega t \qquad (14.12)
$$

This may be obtained from integration of the parabolic velocity expression in Equations 14.5 and 14.6. Thus the pressure vs. flow rate relation exactly parallels the electrical relation:

$$
\Delta p = Q_{ripple}\left[\left(\frac{8\mu\ell}{\pi r_o^4}\right)\sin\omega t + \left(\frac{\rho\ell}{\pi r_o^2}\right)\omega\cos\omega t\right] \qquad (14.13)
$$

Equation 14.13 allows the values of fluid resistance and inductance to be read out directly from comparison to Equation 14.4:

$$R = \frac{8\mu\ell}{\pi r_o^4} \qquad\qquad L = \frac{\rho\ell}{\pi r_o^2}$$

It is, perhaps, serendipitous that the inductance of fluid with a parabolic velocity distribution is the same as that calculated for a fluid with plug flow.

Impedance: As a last topic in this development section, consider impedance. The term impedance is used for the effect of resistors, capacitors, and inductors when used with constant amplitude sinusoidal potentials and currents in electrical circuits. Capacitors and inductors impede current flow, but they do not convert electrical energy into heat. We have already observed the impedance of a resistance, R, and an inductance, $L\omega$, when used in a circuit excited by a constant amplitude sinusoidal potential. Consider the definition of a capacitance from Table 14.1 when exposed to a sinusoidal excitation:

$$\frac{dV}{dt} = \frac{1}{C}i$$

$$V_{ripple}\omega\cos\omega t = \frac{1}{C}i_{ripple}\sin\omega t \qquad (14.14)$$

$$V_{ripple}\cos\omega t = \frac{1}{C\omega}i_{ripple}\sin\omega t \qquad (14.15)$$

Thus Equation 14.15 shows that the impedance of a capacitance is $1/C\omega$. In summary, the impedance forms for constant amplitude sinusoidal excitation are:

RESISTIVE IMPEDANCE	CAPACITIVE IMPEDANCE	INDUCTIVE IMPEDANCE
R	$\frac{1}{C\omega}$	$L\omega$

14.3.2   Example: Impedance

Consider a pump with the characteristics shown in Table 14.2. The pipe diameter has been chosen larger than might strictly be used to ensure that full flow is laminar. This is a point that should be noted: noise control deals with ripple superposed on a base flow. It is the base flow, however, that will determine the flow regime. The value of resistance, $R = 8\mu\ell/\pi r^4$, shown in Table 14.1 is based on laminar flow where the Q value is the ripple flow rate. The ripple flow will not be laminar if the base flow is turbulent.

Table 14.2: Circuit characteristics for worked example of impedance

Characteristic	Value	Units
Pump base flow	80.0	L/min
Number of pump pistons	9	
Pump speed	1800	rpm
Ripple flow amplitude, Q_{ripple}	1.0	L/min
Pipe diameter, d	50.0	mm
Pipe radius, r_o	25.0	mm
Pipe length, ℓ	2.0	m
Absolute viscosity, μ	12.3E−3	Pa · s
Bulk modulus, β	1.2	GPa
Density, ρ	860	kg/m^3

Calculate the pipe velocity for full base flow:

$$v_{base} = \frac{Q_{base}}{\pi d^2/4} = \frac{80\text{E}{-}3 \text{ m}^3/\text{min} \times 4}{\pi(50.0\text{E}{-}3)^2 \text{ m} \times 60 \text{ s/min}} = 0.679 \text{ m/s}$$

Now calculate the Reynolds number for the base flow:

$$Re = \frac{\rho v_{base} d}{\mu} = \frac{860 \text{ kg/m}^3 \times 0.679 \text{ m/s} \times 50.0\text{E}{-}3 \text{ m}}{12.3\text{E}{-}3 \text{ Pa} \cdot \text{s}} = 2374$$

As shown in Section 9.3.1, a nine piston pump will have a fundamental frequency of 540 Hz. In terms of radians/second, $\omega = 3393$ rad/s. We shall now calculate each of the impedances based on this value of ω and the ripple amplitude, $Q_{ripple} = 1$ L/s.

Find the impedance of the resistance, R:

$$\begin{aligned} R &= \frac{8\mu\ell}{\pi r_o^4} = \frac{8 \times 12.3\text{E}{-}3 \text{ Pa} \cdot \text{s} \times 2.0 \text{ m}}{\pi(25\text{E}{-}3)^4 \text{ m}^4} \\ &= 160.4\text{E}{+}3 \text{ Pa} \cdot \text{s/m}^3 \end{aligned}$$

It is probably more meaningful to express this quantity as a pressure drop across the resistance:

$$\Delta p_R = RQ_{ripple} = 160.4\text{E}{+}3 \text{ Pa} \cdot \text{s/m}^3 \times 1.0\text{E}{-}3 \text{ m}^s/\text{s} = 160.4 \text{ Pa}$$

Find the impedance of the capacitance, C:

$$\frac{1}{C\omega} = \frac{1}{(\pi r_0^2 \ell/\beta)\omega} = \frac{1}{(\pi(25.0\text{E}{-}3)^2 \text{ m}^2 \times 2.0 \text{ m}/1.2\text{E}{+}9 \text{ Pa})3393 \text{ rad/s}}$$
$$= 90.1\text{E}{+}6 \text{ Pa} \cdot \text{s/m}^3$$

Express as a pressure drop:

$$\Delta p_C = \frac{1}{C\omega} = 90.1\text{E}{+}6 \text{ Pa} \cdot \text{s/m}^3 \times 1.0\text{E}{-}3 \text{ m}^3/\text{s} = 90.1\text{E}{+}3 \text{ Pa}$$

Find the impedance of the inductance, L:

$$L\omega = \frac{\rho\ell}{\pi r_o^2}\omega = \frac{860 \text{ kg/m}^3 \; 2.0 \text{ m}}{\pi(25.0\text{E}{-}3)^2 \text{ m}^2}3393 \text{ rad/s}$$
$$= 2.97\text{E}{+}9 \text{ kg/m}^4.\text{s} = 2.97\text{E}{+}9 \text{ Pa} \cdot \text{s/m}^3$$

Express as a pressure drop:

$$\Delta p_L = L\omega Q_{ripple} = 2.97\text{E}{+}9 \text{ Pa} \cdot \text{s/m}^3 \times 1.0\text{E}{-}3 \text{ m}^3/\text{s} = 2.97\text{E}{+}6 \text{ Pa}$$

Comments: The form of noise filtering using resistance, capacitance, and inductance is directly akin to filtering out electric AC ripple from DC power supplies [12]. In the past, electrical engineers would use a π filter consisting of two capacitors to ground either side of a series inductance (called a *choke* because of its function). All the components were considered necessary because the frequency being filtered out was quite low, either 60 or 120 Hz in the U.S. The capacitors to ground were selected for low impedance and the choke for high impedance. The fluid power circuit only has one conductor, but the capacitance unit, the large volume added locally to a pipe, is effectively a shunt to ground for the ripple component and should be designed for as *low* an impedance as possible.

In the worked example, just a length of pipe was analyzed. We may conclude that resistance probably plays only a small role in removing ripple. If the resistance has too great an impedance, this will affect the main flow though the system adversely.

On the other hand, the inductive pressure drop for the ripple component, $L\omega$, is quite large and would have little effect on the operation of the circuit for moving loads. This statement will not be entirely correct. Adding inductance will affect the dynamics of a circuit, although this effect was not addressed in Chapter 4.

The capacitive impedance of the pipe section has a relatively high value and could be reduced by introducing a larger volume.

14.3.3   Development of a Lumped Parameter Model

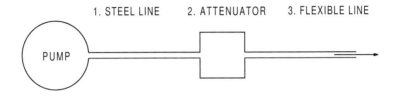

Figure 14.3: Arrangement of pump, flow lines, and attenuator.

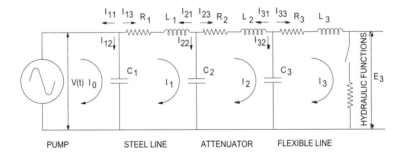

Figure 14.4: AC circuit analogy for system shown in Figure 14.3.

Consider the basic configuration shown in Figure 14.3. A pump feeds into a steel line followed by an attenuator volume and a length of flexible hose before the oil is delivered to the major components of the hydraulic circuit. Length, diameter, and bulk modulus for each element is listed in Table 14.3.

Figure 14.4 illustrates a typical lumped parameter model represented with the use of the electric-hydraulic analogy [13]. The network is the result of applying the lumped parameter approach to the hydraulic tube

Table 14.3: Characteristics with applicable units needed to establish
a model for the system in Figure 14.3

Characteristic	Description	Value	Units
β_1, β_2, β_3	Bulk modulus	1.17, 1.17, 0.83	GN/m^2
μ	Absolute viscosity	0.0147	Ns/m^2
ρ	Oil mass density	856	kg/m^3
ℓ_1, ℓ_2, ℓ_3	Line length	0.53, 0.146, 0.381	m
d_1, d_2, d_3	Tube diameter	0.0079, 0.0762, 0.0079	m
C	Fluid capacitance		m^5/N
R	Fluid resistance		$N \cdot s/m^5$
L	Fluid inductance		$N \cdot s^2/m^5$
I	Fluid flow (current)		m^3/s
E	Fluid pressure (potential)		N/m^2
$V(t)$	Applied potential	2.07E+6	N/m^2
p	Pressure		N/m^2
r	Tube radius		m
Q	Fluid flow rate		m^3/s

arrangement shown in Figure 14.3. The figure shows a volume in the output line from a typical hydraulic pump used in mobile and industrial fluid power systems. Use of the modeling approach described here, when applied to this type of hydraulic line arrangement, will provide the desired attenuation of pump pressure waves in many applications.

The physical system shown in Figure 14.3 consists of a hydraulic pump with an output line that has a volume added to it for use as a pressure wave attenuator. The pump output flow therefore travels through a series of three sections that can be modeled with loops as shown in Figure 14.4. Each section contributes a C, L, and R term to allow formation of the circuit loop equations that follow:

$$V(t) = V \sin \omega t$$

$$V(t) = (1/C_1) \int_0^t (I_0 - I_1)dt$$

$$(1/C_1) \int_0^t (I_1 - I_0)dt + R_1 I_1 + L_1 \dot{I}_1 + (1/C_1) \int_0^t (I_1 - I_2)dt = 0$$

$$(1/C_2) \int_0^t (I_1 - I_2)dt + R_2 I_2 + L_2 \dot{I}_2 + (1/C_3) \int_0^t (I_2 - I_3)dt = 0$$

Differentiation of these equations provides time dependent relationships:

$$\dot{V}(t) = V\omega \cos \omega t$$

$$I_0 = C_1 \dot{V}(t) + I_1$$

$$\ddot{I}_1 = (1/L_1)[-(1/C_2)(I_1 - I_2) - R_1 \dot{I}_1 - (1/C_1)(I_1 - I_0)]$$

$$\ddot{I}_2 = (1/L_2)[-(1/C_3)(I_2 - I_3) - R_2 \dot{I}_2 - (1/C_2)(I_2 - I_1)]$$

In order to complete the solution, the circuit node voltage equations are necessary:

$$I_{21} + I_{23} + C_2 \dot{E}_2 = 0$$

$$I_{31} + I_{33} + C_3 \dot{E}_3 = 0$$

These equations are most useful when arranged as:

$$\dot{E}_2 = (1/C_2)(-I_{21} - I_{23})$$

$$\dot{E}_3 = (1/C_3)(-I_{31} - I_{33})$$

These two equations may be combined with the four differentiated loop equations to provide solutions for E values and I. The equations may be simplified by noting that current values I_{21}, I_{23}, I_{31}, and I_{33} are equal to $-I_1$, I_2, $-I_2$, and 0 respectively.

These modeling equations, when used with the electric-hydraulic analogy terms noted above, provide a model for the selection of an attenuator volume for a particular fluid pump's output line. A variety of parameter combinations can be examined quickly with use of the mathematical model and appropriate computer software [6, 7]. A similar analysis for fluid transients is described by Wylie and Streeter [14].

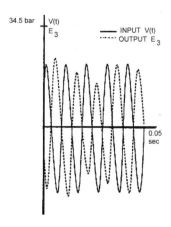

Figure 14.5: Plot for 100 Hz, 2070 kPa amplitude input pressure wave.

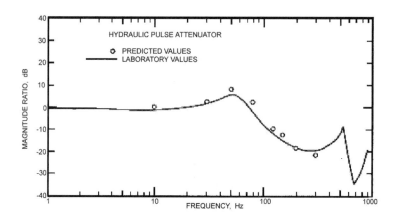

Figure 14.6: Plot of magnitude ratio versus frequency for attenuator.

14.3.4   Example: Curing Noise from Tractor Hydraulics

A variety of applications have been tried with the approach described here. The information shown in Figures 14.3 through 14.6 applies to a medium-size agricultural tractor hydraulic system. Once the mathematical model was developed, this led to the satisfactory addition of an attenuator to this line of machines. Noise and system instability were reduced noticeably with the use of a properly sized attenuator.

The pump output line and attenuator were made of steel for this system. Dimensions of the parts are shown in Table 14.3. The output line from the attenuator was made of rubber and steel braid construction. Flexible lines expand and contract under pressure. Therefore, a lower bulk modulus, β, is used in this section of the model, since the flexible line characteristics cause a lower functional value. This reduced value is sometimes referred to as the apparent or effective bulk modulus (Chapter 2). Values of the apparent bulk modulus depend on the length and diameter of the line.

In order to develop a successful attenuator volume, the appropriate physical dimensions were applied in the mathematical model. The model was then solved with computer simulation software. In addition to the geometric dimensions, other quantities used in the model are noted in the table. The mathematical model was run at various frequencies that correspond to actual pump speeds. At each frequency, a plot was produced, such as that shown in Figure 14.5. The pump input pressure wave amplitude was 20.7 bar as shown on the plot. The other line plotted is the attenuated pressure wave from the output line on the attenuator volume. In Figure 14.5, the magnitude ratio of the pump output pressure wave and the pressure wave from the final line section is plotted versus the pump piston frequency. The relationship for magnitude ratio in decibels, dB, is obtained here as:

$$A_{dB} = 20 \log_{10} \frac{E_3}{20.7} = 20 \log_{10} E_3 - 26.32$$

The pump under study had eight pistons in a radial configuration (Figure 8.5). Inspection of Figure 9.12 shows that an eight piston pump will have eight volume pulsations per revolution, this will result in eight pressure cycles per revolution. When this number of pressure cycles per revolution is multiplied by 2250 rpm, the pump shaft speed, the resulting frequency is 300 Hz. The magnitude ratio is a negative number at this frequency. A wide band of frequencies in the primary operating area of pump speeds have negative magnitude ratios. The attenuator volume, therefore, is well sized for this particular pump and its output line configuration. The circled points on Figure 14.6 are the result of the analysis from the mathematical model.

If phase angles are of interest, they may be deduced directly from the plots, such as those shown in Figure 14.5. Phase angles of about 180° are produced near the primary operating frequency of 300 Hz.

The simulation was run without output flow from the pump. The pump pressure wave was used as the input function. This approach allows a satisfactory study of the attenuator size selection and effectiveness. Parts made with dimensions used in the model were tested in a laboratory arrangement that duplicated the simulation operation. The solid curve shown in Figure 14.6 is the result of the laboratory testing.

The agreement between the simulation results and the laboratory data, for this particular arrangement, is very good. The results confirm that the somewhat elementary model will give good usable results in the selection of a satisfactory attenuator volume. As shown in Figure 14.6, a significant reduction of 20 dB is realized at the pump rated speed of 300 Hz.

Laboratory data, as recorded by the solid line in Figure 14.6, was measured with the output transducer at the end of the attenuator output line. Pressure waves were imposed on the system, by the pump, without actual flow through the lines. Laboratory and simulated data, for tests with actual flow, were quite similar to tests with no actual flow. Predominant natural frequencies appear for the system at about 55 and 550 Hz. An increase in flow through the system did not affect the natural frequencies, however, amplitude at the natural frequencies was somewhat reduced. Amplitudes at other frequencies were also slightly reduced.

14.4   EFFECT OF ENTRAINED AIR IN FLUID

Examination of the equations that provide a mathematical model for an attenuator pulsation damping system show the importance of the bulk modulus term β. Selection and maintenance of the fluid in use, therefore, becomes important. A petroleum-based fluid can absorb about 9% of air by volume at room temperature and atmospheric pressure [8].

It is important that all air be bled from a fluid power system before operation. Otherwise, the air can become entrained in the fluid and cause a change in the value of the bulk modulus (Chapter 2). The analysis presented here illustrates another major effect that bulk modulus has on fluid power systems. In addition to the change in the value of the bulk modulus, entrained air can cause other undesirable effects in a fluid power system. Considerable noise can be generated when air bubbles are released within a high-pressure fluid power stream. Cold fluid will hold more air in solution than hot fluid. Also, the amount of air entrained in the fluid is inversely proportional to the fluid pressure. Along with the selection of an appropriate attenuator volume, or the use of other noise suppression components, it becomes important to keep in mind techniques to minimize entrained air in the fluid.

14.5 FURTHER DISCUSSION OF THE MATHEMATICAL MODEL

Additional work with the type of mathematical model presented here will likely bring improvements in the development of attenuators. The preliminary work that has been done produced the conclusions that follow:

1. The general theoretical modeling techniques developed in this text provide useful tools for sizing a pressure wave attenuator.

2. Results of a typical model, when run with a computer simulation program, produce results that closely approximate actual laboratory data.

3. Information recorded in Figures 14.3 to 14.6 represents work on a pre-production system that performed satisfactorily when applied to a line of agricultural tractor hydraulic systems.

4. Several particular influences were noted in laboratory and simulation results. The items that follow refer to a horizontal shift of the output curve relative to the input curve:

   (a) A reduction in oil bulk modulus β caused the output curve to shift to the left.

   (b) An increase in the length of the line, which feeds into the attenuator volume, caused the output curve to shift to the left.

   (c) An increase in the attenuator in input line diameter caused the output curve to shift to the right and reduced the effectiveness of the volume as a pulsation damper.

   (d) A reduction in attenuator volume caused the output curve to shift to the right and reduced the effectiveness of the volume as an attenuator.

14.6 OTHER METHODS OF NOISE CONTROL

Various other means of attenuating pressure waves are in use in fluid systems. The fluid volume may not always be the best for a particular system. Also, sometimes several means are necessary to obtain desired results. Therefore, a brief discussion is being included of the attenuation devices

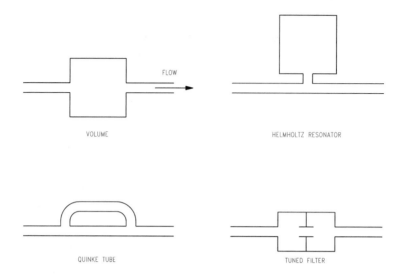

Figure 14.7: Some common types of fluid pressure wave attenuators.

shown in Figure 14.7. Generally, the tuned filter, the Quinke Tube, the Helmholtz resonator, and the accumulator are of value as attenuators [2-4, 8, 15]

Tuned filters are sometimes used with a combination of devices that broaden the tuning over a wider range. This occurs because tuned filters can provide high attenuation, which applies only in narrow bands. A tuned filter that is effective for a particular pump speed is of no value if the pump speed must be changed, so additional means of attenuation becomes necessary.

The Quinke Tube is an arrangement that splits flow equally between two lines of analytically selected lengths. The two flows are then recombined at a downstream junction. The device works because the two flows are out of phase when merged because they travel different distances. A fundamental frequency and its harmonics can be cancelled out with this type of line arrangement.

The Helmholtz resonator principle can be used by adding a volume in a side branch as shown in Figure 14.7. The fluid in the volume and its connecting line form a resonant subsystem. Only pulsation waves at frequencies below the particular resonant frequency of the resonator can be attenuated by this arrangement. No useful attenuation effect is gained at higher frequencies

14.7 DAMPING METHODS

Reduction in the undesirable oscillations of valves has been found to be a very useful means of reducing noise and vibration. Many types of damping effects have been applied to valves. The information that follows outlines a very successful method of valve oscillation damping. The damper was applied to a braking system valve on an earth-moving machine and later adapted to other similar valves [16].

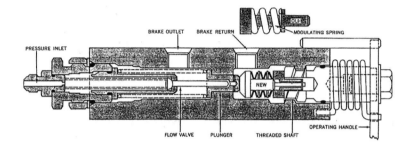

Figure 14.8: Cross-section of brake operating valve.

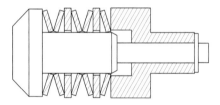

Figure 14.9: Cross-section of disc spring assembly.

The hand-operated brake valve shown in Figure 14.8 energizes the brakes on a tractor-towed, earth-moving machine. The valve is mounted in the steering column on the tractor and is actuated by a lever that rotates a threaded shaft. The shaft compresses the pressure modulating spring, which then depresses a plunger and flow valve letting oil discharge through the brake outlet port.

Audible chatter not only annoyed the tractor operator, but the vibration also caused repeated failures of the valve inlet line. The valve vibrated at two frequencies: one at 120 Hz, the other at over 1000 Hz. Both frequencies were independent of supply pump pulsations and seemed to originate in the brake valve.

Laboratory tests proved that the pressure modulating spring was the culprit. The original spring was a conventional steel coil spring. Use of a solid spacer in place of the spring stopped the vibration, however, the modulating feature supplied by the spring was then lost. Various dampers were considered with the spring in an effort to replace the modulating feature. Laboratory trials with the friction damped spring pack, shown in Figure 14.9, appeared to be the best solution.

The pack has ten disc springs, stacked in a series-parallel arrangement. This provides a spring constant approximately equal to that provided by the steel coil spring. Two flat washers were added to provide friction.

The spring pack modulates satisfactorily. There was enough friction damping to eliminate noise on field and laboratory test units. Inlet line failures due to vibration were eliminated. Hysteresis was evident in the operating characteristics of the valve. Laboratory tests determined the value of the spring pack friction force. The load-deflection curve for the pack appeared as a typical hysteresis loop. The friction damping force varies with spring deflection and becomes greater with increased deflection. The hysteresis was noticeable to the operator, but was not large enough to defeat using the self-damping the spring pack provides.

To simplify valve assembly and service procedures, springs and washers were assembled on a retainer and guide and secured with a retaining ring. The principle involved in the damped spring pack was applied to other valves and proved useful in quieting similar noisy units.

PROBLEMS

14.1 Consider the oil filled pipe examined in Section 14.3.2. Let one end of the pipe be connected to a source of sinusoidal flow. In this configuration, the capacitance is in parallel with the series connection of the resistance and the inductance. In the electrical analogy, this would be known as a resonant circuit because there is one frequency at which the voltage across the circuit would be a maximum. Write an equation for the pressure *across* the circuit when a time varying flow rate is applied at one end. Work in symbols.

Hint: Apply the varying potential to the capacitive leg of the circuit in terms of the varying flow rate. Do the same for the resistance and inductance leg. Apply the Laplace transform to the two equations individually. Apply Kirchhoff's law to the currents and combine the two Laplace equations and rearrange so the pressure equals a transfer function of the R, C, and L components in the circuit multiplied by the flow rate as an input function. At this stage return to Problem 6.2 and examine the block diagram in light of your analysis.

Review the material on a spring-mass-damper presented in Chapter 5 and write an expression for the damped frequency at resonance. Reorganize the transfer function denominator in terms of ζ and ω_n.

14.2 Use the values obtain for R, C, and L in Section 14.3.2 to calculate the damped resonant frequency for that particular pipe. Use a program that can draw frequency response diagrams to draw an amplitude ratio curve covering frequencies from $(1/100)\times$ to $100\times$ the damped resonant frequency. Comment on the suitability of the pipe to provide ripple damping for a five cylinder pump operating at 1800 rpm.

14.3 Assume that the pipe model introduced in Problem 14.1 is valid for initial design of a noise attenuator. Design a noise attenuator for a seven piston pump running at 3000 rpm that has an attenuation of 20 dB at the fundamental frequency. The pump delivers 20 L/min. Flow in the inductive element is to be laminar. An additional inductive element can be added ahead of the inductive/resistive element. The capacitive element may be designed separately from the inductive element, but the capacitance of the oil in the inductive element should be included in the calculations. There is no one best answer for this problem. Ensure that you consider any resonant frequency effects and keep these at a speed at which the pump will only operate transiently. The final design should consider the resistance element with respect to the increased pressure response at resonance. Use β_e = 1.2E+9.

REFERENCES

1. Ezekiel, F. D., 1958, "The Effect of Conduit Dynamics on Control-Valve Stability", Transactions of the ASME, May, pp. 904-908.

2. Vander Molen, G., and Akers A., 1986, "Resonance of a Pressure Control Valve Sensing System", Paper 86-WA/DSC-29, American Society of Mechanical Engineers, pp. 12.

3. Henke, R. W., 1986, "How to Reduce Noise in Hydraulic Systems", Hydraulic and Pneumatics, March, p. 58.

4. Miller, J. E., 1973, "Silencing the Noisy Hydraulic System", Machine Design, June, p. 138.

5. Fawcett Engineering Limited, 1986, "Fawcett Hydraulic Noise Attenuators", Fawcett Engineering Limited, Dock Road South, Bromborough, Merseyside L62 4SW, U.K.

6. M. P. Gassman, 1986, "Noise and Pulsation Control with Pressure Wave Attenuators", Proceedings of the 7th International Fluid Power Symposium, BHRA, Bath, U.K.

7. Gassman, M. P., 1987, "Noise and Pressure Wave Control with Hydraulic Attenuators", SAE Paper 871682, Milwaukee, WI.

8. S. J. Skaistis, 1979, "New Techniques Muffle Hydraulic Noise", Machine Design, March, pp. 120-126.

9. Foster, K. and Parker, G. A., 1964-1965, "Transmission of Power by Sinusoidal Wave Motion through Hydraulic Oil in a Uniform Pipe", Proceedings, Institution of Mechanical Engineers, **179 Pt.** **1**, (19), pp. 599-614.

10. Krishnaiyer, R., 1968. "Fluidic Transmission Line Properties", Instruments and Control Systems, October, pp. 93-95.

11. Oldenburger, R., and Goodsen, R. E., 1964, "Simplification of Hydraulic Line Dynamics by Use of Infinite Products", Journal of Basic Engineering, March, pp. 1-10.

12. Brophy, J. J., 1977, *Basic Electronics for Scientists*, 3rd ed. McGraw-Hill Book Company, New York, NY.

13. R. E. Raymond, 1961, "Electrohydraulic Analogies, Part 1", Hydraulics and Pneumatics, March, pp. 65-69.

14. Wylie, E. B. and Streeter, V. L. with Suo, L., 1993, *Fluid Transients in Systems*, Prentice Hall, Englewood Cliffs, NJ.

15. Flippo, W. T., 1986, "Sharp EL512 Calculator Helps Size Accumulators, Part 3 Pulsation Damping", Hydraulics and Pneumatics, January, p. 139.

16. Gassman, M. P., 1968, "Quieting a Noisy Valve", Hydraulics and Pneumatics, April, p. 131.

INDEX